Megacities
The New Global Community

Wendy Slone

Kendall Hunt
publishing company

Kendall Hunt
publishing company

www.kendallhunt.com
Send all inquiries to:
4050 Westmark Drive
Dubuque, IA 52004-1840

ISBN 978-1-4652-4195-5

Printed in the United States of America
10 9 8 7 6 5 4 3 2 1

Contents

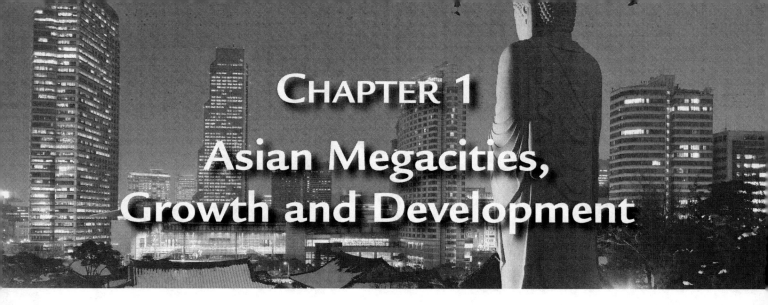

CHAPTER 1
Asian Megacities, Growth and Development

Risks and Opportunities of Urbanization and Megacities

Theo Kötter

URBANISATION AND MEGACITIES—
THE CHALLENGE OF THE 21ST CENTURY

The 21st century is the century of the cities and of urbanisation (Hall/Pfeiffer 2001). According to The State of World Population 2001, an actual report from the United Nations Population Fund, roughly 2.8 billion people live already in cities and by 2015, that number will have risen to 3.9 billion. The total population is increasing by 280.000 people per day. Nearly three-quarters of them will be inhabitants of the developing world. While in developed countries urbanisation has mainly taken place in the second half of the 19th century, developing countries are in the middle of their urban growth now. In Europe already 76 % of the population live in cities. Urbanisation has come to stand still and we can notice a process of dis-urbanisation and sub-urbanisation caused by a high rate of motorisation combined with prosperity and the development of traffic and communication infrastructure.

Compared to this in the developing countries, the urbanisation is increasing rapidly and will continue during the next decades. For the first time in 2007 more people live in cities than in the rural areas. The highest growth will occur mainly in the cities of Asia and Africa, that are now more than two-third rural, will be half urban by 2025. Never before urban population has expanded so fast because of the progress in agriculture, science of nutrition and medicine. For example Dar es Salam, Tanzania, has a growth rate of 6% per year, which leads to a doubling of population every 13 years. A high birthrate combined with an increasing migration from the rural areas that is reinforced by the

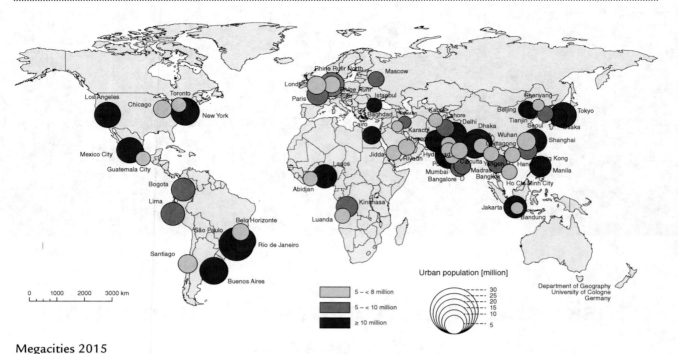

Megacities 2015

Source: http://www.megacities.uni-koeln.de/documentation/megacity/map/MC-2015-PGM.jpg

so called "push-factors" (unemployment, low standards of housing and infrastructure, lack of educational facilities) and "pull-factors" (economical opportunities, attractive jobs, better education, modern lifestyle) lead to the very dynamic growth process. Most of this growth is taking place in the poor quarters of the cities. One can imagine the challenges to manage cities in a sustainable manner when their population doubles every 13 years.

The number of megacities, which have 10 million or more residents, is increasing worldwide: 1950: 4, 1980: 28, 2002: 39, 2015: 59. Two third of them are situated in developing countries, especially in South-East-Asia. In 2002 already 394 million people live in megacities, 246 million of them in developing countries, more than 215 million in Asia. In the year 2015 the total population of megacities worldwide will be about 604 million and the further rate of growth will be high, as the development of Mexico City, São Paulo, Seoul, Bombay, Jakarta and Teheran shows which population has tripled between 1970-2000. According to the estimation of the UN concerning the number of megacities in 2015, Bombay (22.6 mill. inhabitants), Dhaka (22.8), Sao Paulo (21.2), Delhi (20.9) and Mexico City (20.4) will be five of the worldwide six biggest megacities each with much more than 20 million inhabitants. 100 years ago London (6.5) has been the greatest city (one million more inhabitants than New York), today it is a shrinking town.

The rapid process of urbanisation and the growing number of the megacities, cause a lot of different ecological, economical and social problems and risks. This impacts cause challenges for urban policies and urban planning strategies to manage the development in a sustainable way, when the population in some cities doubles every 10 to 15 years.

The reason that the agglomeration and metropolitan areas as well as and megacities come into the international focus of policy and science are their serious impacts on the global environment like the enormous land consumption, air pollution, water scarcity, poverty, social segregation and vulnerability. As the numerous national and international conferences on urbanisation and megacities show, there is an obvious need for more and better urban development strategies, long term land policy and forceful urban management.

EFFECTS AND IMPACTS OF URBANISATION AND MEGACITIES

The following characteristic of megacities has to be mentioned as the typical features that bring these agglomerations into the focus of science, policy and economy. These characteristics imply a lot of serious risks, but also potentials and opportunities for the regional and global development:

- **Density:** Megacities show the highest density of inhabitants, industrial assets and production, social and technical infrastructure. Metropolitan areas and especially megacities become more and more the centres and nodal points of the global economy. With their important role as centres of political and economic decisions they are promoters of national and international developments. Furthermore in this areas lots of highly qualified and "inexpensive" skilled labour are available and also the concentration of capital stock make them attractive for investments. Urban Agglomerations and megacities generate a lot of income and their local economies have an importance for their rural surroundings. It is no coincidence that in megacities worldwide have increasing part of the GND (e.g. Thailand: 20% of the population lives in Bangkok, but 70% of the GND).

- **Dynamic:** Megacities are characterised by highest dynamic in the fields of spatial and demographic growth, change of land use and consumption of land for settlement purposes, that mostly take place in absence of urban planning and on foreign plots. Also the formal and informal urban economic sectors are on a high dynamic level. The local, regional and global markets and the connection with the international economic circulation induce various increasing economic activities, so that megacities have the economic potentials and power to initiate economic growth also in the regions around the urban areas.

- **Settlement, infrastructure and land tenure:** In the most agglomerations and megacities urban planning and public infrastructure can guide the urban development in order to achieve a proper sustainable structure only very partially. The extension of cities is always in advance of urban development work and the provision of public facilities. Beside the "proper city", which is in the focus of strictly conventional urban planning, all the other quarters and districts of the agglomeration and megacities grow up outside the law and without the benefits of urban planning. Especially the informal housing areas and in many times also illegal housing areas (squatters), that are build up by the migrants themselves lead to an extensive settlement structure. The illegality of those residential areas results mainly from the land tenure system.

- In many cases there is a lack of an efficient infrastructure for the public and private traffic, of proper garbage removal and also of sewage systems with wastewater purification. Most city-dwellers have no sanitation facilities and rainwater drainage systems are totally inadequate. This situation has serious consequences on the environment and public health. About 1.5 billion people live in slums and squatters.

- **Socio-economic disparities:** In megacities we can recognise a wide range of social standards and social fragmentation, social-cultural conflicts because of the different background of the immigrants and a great number of urban poor, which are bad provided with public facilities and infrastructure and their housing areas are often edged out by stronger economic purposes and land use. The development and extension of cities is accompanied with mounting urban poverty. Roughly a quarter of the population of the developing countries (1.3 billion people) are living in situations of absolute poverty on less than one dollar per day (UNDP 1997). A resident in a poorer housing area in Chicago has better living conditions than about 80 % of the megacity-dwellers in the developing countries. E.g. in Calcutta, Madras, Bombay and Delhi more than 50 % of the inhabitants are living in squatters. The growing socio-economic disparity within the megacities and the lack of social cohesion is the most serious explosive charge.

- **Risks and vulnerability:** Considering the density and number of inhabitants and also the accelerated development megacities run highest risk in the cases of men made and natural disasters. Related to the population we have a high rate of consumption of natural resources especially land for new settlement areas, water and energy. The result is an ecological strain of the environment with serious pollution of the air, water and soil. Air pollution, mainly caused by traffic, traffic congestion and industrial production, in most cities of developing countries exceeds the environmental standards considerable. The annually losses of the world's GND determined by air pollution is estimated between 0.5 and 2.5 %. Another considerable problem is the provision of the residents with clean drinking water.
- At many times the location of new squatters of rapid growing cities and agglomerations is not suitable according to a proper and safe urban development. The main reason is, that e.g. in the 1990's, 60 to 70 % of urbanisation was unplanned, often in areas, which are adjacent to industrial zones, known to be highly seismic or flood prone. The accelerated and uncontrolled growth has contributed to the ecological transformation of the cities and their immediate surroundings (pressure on scarce and sealing increase the volume and speed of rainfall runoff that make many cities more vulnerable to flash floods). Furthermore other factors depending on the local circumstances contribute to the urban vulnerability: lowering or rising water table, rising sea level, earthquakes, storms and land slides. Through lack of choice the ongoing urbanisation forces more and more populations to settle on those disaster prone areas.
- **Governance:** One of the greatest challenges of agglomerations and megacities is their governability and one can recognise a crisis of urban government in this. The experiences show that the possibilities of traditional forms of centralised governance with top down strategies are restricted because of the extension, highly dynamic and highly complex interactions within the megacities and also with their surroundings. In the case of spatial planning, decentralisation and innovative planning processes with intensive participation are necessary.

All in all agglomerations and megacities are not only risk areas of the global change, but they also provide opportunities. They are the engines of the economy and in many cases, they are also precursors of the urbanisation.

APPROACHES AND PRINCIPLES FOR THE DEVELOPMENT OF AGGLOMERATIONS AND MEGACITIES

Models of Sustainable Development

The development of megacities and sustainability seems to be contrasts, that cannot go together at the same time. The high rates of land and energy consumption, the severe pollution of air, water and soil at present and the ongoing social fragmentation are not in compliance with the aims of a sustainable development. To cope this risks and challenges, considering the undamped growth, a spatial concept with a decentralized structure should be underlied that includes the urban and the surrounding rural areas. In the past, different models of sustainable development have been discussed, but there is no general admitted structure, that solves the risks of megacities. With view on megacities and agglomerations a regional settlement structure has to be designed which set up on the elements density, mixing of different land uses, polycentrality and capacity of public mass transport systems and public facilities. These are the prerequisites for achieving the ecological, social and economic targets of sustainability.

The priority must be to slow down the urban growth. Therefore the living conditions and the economic basis in the rural areas must be strengthened, to prompt the inhabitants to stay there. Therefore it's a vital necessity to promote new forms of cooperation between cities and between the cities and the villages at the regional level.

Strategies of Urban Development

To achieve a proper development of agglomerations and megacities a comprehensive plan is indispensable, that provides guidelines and principle goals for the urban development as well as for the development of the and that also provides the basis for construction immediate plans for economic and social development, area plans, district plans, detailed plans etc. In accordance with the sustainability, the integration and coordination of urban and rural areas with the central city should be a main principle. This requires a "multi-center", "multi-axis" and "multi-level" urban spatial structure. For example the comprehensive plan of Shanghai (1999–2020) lines out five levels that refers to five scales. The urban system is composed of the Central City, New Cities, Central Towns and the Ordinary Towns and Central Villages.

In case of the urban development of megacities a shift of urban policy and also of planning strategies is fundamental. This includes a legalisation and registration of informal settlements slums and squatters. Furthermore considerable social improvements and an access to schools and other educational institutions are necessary. Self-help housing improvements must be strengthened combined with the access to land to enhance the living condition, the identification with the quarter and at least the engagement for the (local) community.

The final declaration of the Heads of State and Government and the official delegations from the countries attending the 2nd United Nations Conference on Human Settlements, Habitat 11, held in June 1996 in Istanbul, proclaimed the "right to adequate shelter for all" as one of the key themes of the conference. A billion people are today without a decent home and a hundred million are completely homeless. This gives a measure of the needs and the singular importance of the housing problem. Access to housing is now recognized as being central to social cohesion and a key factor for development.

Long-term Land Use and Land Management Strategies

A long-term land use and land management strategies need reliable economic conditions and authoritative legal regulations. Therefore the reform of land tax must be discussed considering land policy, fiscal, social and ecological aspects. A sustainable urban development requires to prevent land fragmentation and also social fragmentation. Considering the rapid growth and that 60 to 70 % of the urbanisation are uncontrolled a comprehensive urban planning has to be developed and monitoring system must be established. Therefore the designation and mobilisation of building land is one of the long-term tasks to be addressed by the local authorities.

To improve the housing situation at long-term, first the problems of land management and land use have to be solved. This requires legal instruments for more secure access to land and planning techniques for urban development and facilities. This frame must be provided at the national level by the State on the national level. If an adequate political, legal and institutional frame has been established, civil society can play an enabling role to implement the land policy and land administration.

In practice the greatest challenge is not elaborating a comprehensive plan of the city or regional development, but providing sufficient urban land for housing and other purposes at a reasonable price and also the indispensable technical infrastructure. Urban land manager must be capable of evolving a coherent vision of the cities future and also mobilising private investment both for housing and for urban facilities and services.

Cost and Energy Saving Facilities and Innovative Transport Systems

The provision of infrastructure for the purposes of transport, communications, energy, drinking water, sewage purification and sold waste treatment contribute the economic development, make the territorial areas more competitive and attractive and promote regional economic integration and social cohesion. But the developing countries cannot support their cities in this fundamental tasks, because they have to cope with severe, long-term budgetary problems. That's why there will be a widening gap between the growing demand and the current provisioning of water and sanitation in the megacities with serious problems for the heath of the residents. The current financial gap is estimated to be US $ 16 billion a year. Especially public-private partnerships can bring efficiency gains and cost-effectiveness in the water sector.

To influence city-dwellers' living conditions and economic development the public authorities have to be involved in producing and managing technical urban infrastructure facilities and services such as roads, transport, electricity, telecommunications, water, sanitation and waste treatment and also social facilities and services in the strategic fields of education and health. In megacities and agglomerations of the developing world there is considerable leeway to be made up and it will take a long time to achieve this with the 200 billion dollars invested each year by developing countries (4 % of their national product). E.g. only the needs of India have been estimated at 50 billion US $ per year. The main problem is to mobilise new external resources to finance gradual improvements of the urban infrastructure. Funds for new infrastructure are required and also for the maintenance and rehabilitation of existing infrastructure to avoid deficiencies. In this fields priorities must be given: Financing and management of existing facilities or investments in future facilities?

The systematic extension of public transport systems into the surrounding is necessary to slow down the migration from the rural areas. A rail transit network with different speed and high capacities, passenger transit pivots and parking lots are important elements of an efficient mass public transport system. E.g. Shanghai has designed an urban transportation plan which consists of high speed rail lines, urban metro lines and urban light railways in order to limit the quantum of cars, motorcycles and powered bicycles. By means of high-tech, the research and development of intelligence transit systems should be forced. This is at the same time a policy reduce energy demand and also the emission of greenhouse gas. But in many cases efficient public mass transport systems are inevitable for these cities.

Good Governance

With the ongoing growth of urban agglomerations and megacities, good governance within the cities become highly complex. One of the main problems in governing megacities and agglomerations is their big extension and high population. These cities have to co-ordinate their activities through local units. To shape policy in a local way it will be necessary to divide megacities and agglomerations in manageable territorial areas and to decentralise some responsibilities to the local actors and initiatives. At the same time it is important to ensure and to organise solidarity between all urban territorial areas and the rural surroundings and the central government. But there is still a need for city or even regional bodies responsible for city-wide or region-wide tasks like mass transit, waste disposal or structural planning.

In many countries decentralisation of urban government is in progress and forced with heavy emphasis. The aim of this comprehensive movement is to improve urban living conditions by addressing needs as directly as possible and to enable city-dwellers to participate in city matters. It is a question of efficiency of administration and also of political strategies that involves reorganising the political authorities and administration responsibilities between the central and the local authorities. In the decentralising process, a balance must be found between internal socio-political concerns and the common development strategy of the megacity.

But decentralisation by its own is not yet a guarantee for good governance. Decentralization requires also capacity building for an efficient local urban management. Inadequate mobilization of local resources is a major obstacle for managers in the performance of their tasks. Local tax levying capacities are poor owing to the lack of any organized collection and control system. Taxation methods are often discretionary and do not encourage taxpayers to comply. House and land tax legislation and tax of urban economic activities tend to be unproductive because they have not kept pace with economic and social development.

This strategy is largely determined by the objectives and requirements of city-economic and budgetary balances, by the land use planning strategy, the financial policy, credit regulations, education and health policy, land and tax legislation. No foreign model of decentralisation is transferable and it is possible for countries to be enriched by other experiences and best practices, but they have to develop their own appropriate model.

CONCLUSIONS

Megacities and urban agglomerations are complex and dynamic systems that reproduce the interactions between socio-economic and environmental processes at a local and global scale. Despite of their importance for economic growth, social well-being and sustainability of present and future generations, urban areas have not received the level of attention they require in the study of global environmental change. The increasing number and extent of recent natural and men made disasters illustrate the devastating consequences of some of the above mentioned trends and impacts. Global environmental change covers a diverse and broad range of issues. Megacities and urban agglomerations are certainly major source for changes in land use and land cover, and they are major users of energy, natural resources and food, but they offer a unique set of opportunities to advance the creation of new conceptual framework for research. Especially an integrative approach of the physical, social and ecological aspects of urban growth on one hand and urban planning and land management on the other hand is missing so far. Interdisciplinary and multidisciplinary perspectives will improve a better understanding of the process of urbanisation and megacities and their governance.

REFERENCES

Bundesamt für Bauwesen und Raumordnung (1999): Urban Future. Preparatory expertises (Overviews) for the Word Report on Urban Future for the Global Conference on the Urban Future URBAN 21, Bonn.

Hall, Peter; Pfeiffer, Ulrich (2001): URBAN 21. Der Expertenbericht zur Zukunft der Städte. Stuttgart, München.

Konrad-Adenauer-Stiftung (2003): Megacities III, Handlungsmodelle und strategische Lösungen. Kongress 24.-26.11.2003, Wesseling.

Magel, Holger; Wehrmann, Babette (2001): Applying Good Governance to Urban Land Management—Why and How?—In: Zeitschrift für Vermessungswesen, Heft 6/2001.

Toepfer, Klaus (2003): Zukunftsbeständige Stadt- und Regionalentwicklung: Leitmotiv für die Problembewältigung der Megacities. Vortrag im Rahmen des Kongresses Megacities III, "Handlungsmodelle und strategische Lösungen" in Wesseling.

CHAPTER 2
Forces that Shaped Asian Megacities—Colonialization, Urbanization and Globalization

Asia's Urban Century—Emerging Trends

Rakesh Mohan

Keynote address by Mr Rakesh Mohan, Deputy Governor of the Reserve Bank of India, at the Conference on Land Policies and Urban Development, organised by the Lincoln Institute of Land Policy, Cambridge, Massachusetts, 5 June 2006.

URBANISATION: A RECENT PHENOMENON

Widespread all pervading urbanisation is a truly twentieth century phenomenon. Although cities have always existed, even cities such as Memphis, Babylon, Thebes, Athens, Sparta, Mohen-ja-daro and Anuradhapura existed in antiquity, there is little evidence of widespread urbanisation in the early years of civilization. Rome was perhaps the first settlement to reach 1 million people in BC; only in 1800 did London become the second city to reach this population size.

In 1800, only 2 per cent of the world's population was urbanised. By the year 1900, out of a total world population of close to 1.5 to 1.7 billion, only 15 percent of the population, about 250 million, lived and worked in urban areas, a number lower than the total urban population of India alone today. By the year 1950 the proportion of urban to total global population had risen to 30 per cent, with Europe, North America and Oceania having the highest levels of urbanisation then. By the year 2000, 2.8 billion people lived in urban areas equaling approximately 49 percent of the world's population. So the pace of urbanisation witnessed in the twentieth century was truly unprecedented, and it is a wonder that the world has coped as well as it has. We are now at a turning point in human history: the number of people living in cities is about to exceed those in the countryside, perhaps in this calendar year.

The last 50 years have been truly remarkable in terms of the number of people who have been successfully absorbed in cities in a time period that is incredibly short by historical standards. While the world's urban population grew by approximately 500 million between 1900 and 1950, it grew by 2.1 billion in the next fifty years; and is expected to grow by a similar magnitude in just the next thirty years. The speed of urbanisation in Latin America in the second half of the twentieth century was spectacular, vaulting from a just over 40 percent urbanisation level to 75 percent by the end of this period which was also a period of rapid population growth and demographic transition. As may be seen (Table 2.1) the focus of change is now in Asia with the urban population expected to double in the next 30 years or so. This phenomenon of such rapid urbanisation is indeed unprecedented and it has changed human geography beyond recognition. In the process the complexion of development objectives and processes has also undergone significant change.

In the last two centuries, cities have consistently provided the environment for institutional and technological innovation, and have often been referred to as "engines of economic growth"; "agents of change" and "incubators of innovation". Between 1960 and 2000 world output went up four fold, while urban population almost tripled, taking the world from 33 percent urban to 47 percent urban in forty years (Table 2.2).

The twenty-first century will therefore be an urban century. For the first time in human history, more people will live in cities than in the countryside. The urban situation will get more pronounced as the century unfolds. As in the last fifty years, developing countries will be urbanising at a much more rapid pace than developed countries.

Table 2.1   Urban Population Growth Across the Globe

Region	Urban Population 1900 in Millions	Urban Population 1900 Per cent of total	Urban P ulation 1950 in Millions	Urban P ulation 1950 Per cent of total	Urban Population 2000 in Millions	Urban Population 2000 Per cent of total	Urban Population 2030 in Millions	Urban Population 2030 Per cent of total
Africa			32	14.7	295	37.2	787	52.9
Asia			244	17.4	1376	37.5	2679	54.1
Latin America & Caribbean			70	41.9	391	75.4	608	84.0
Oceania			8	61.6	23	74.1	32	77.3
Europe			287	52.4	534	73.4	540	80.5
North America			110	63.9	243	77.4	335	84.5
Global Total	~250	~15	751	29.8	2862	47.2	4981	60.2
Increase			501	14.8	2111	17.4	2119	13.0

Source: United Nations (2002).

Table 2.2   Global GDP and Growth in Urban Population

	1960	1970	1980	2000
World GDP (constant 1995 $ trillion)	7.9	13.5	19.5	34.3
Share of agriculture in world GDP	—	—	6.5	3.9
Share of industry in world GDP	—	—	38.0	20.8
Share of services in world GDP	—	—	55.5	66.3
World population (mm)	3020	3675	4428	6053
Per cent urban population	33.3	36.5	39.3	46.7

Source: World Bank database.

A review of the regional dynamics of urbanisation reveals interesting developments (Table 2.3). There has been a dramatic shift of the fulcrum of urban population away from Europe and North America to the developing regions of the world. During the period 1950-2000, the growth rate of urban population in Europe and North America was about 1.5 per cent. The share of Europe and North America in global urban population declined from about 53 per cent in 1950 to 27.5 per cent in 2000 and is expected to decline further to about 17 per cent by 2030. Africa has experienced consistent high growth in its urban population, which grew at an annual rate of 4.4 per cent during 1950–2000, and its share in global urban population is expected to rise to 16 per cent by 2030 (from 4.3 per cent in 1950). Latin America has now become predominantly urban, surpassing urbanisation levels in Europe and will almost be at par with North America by 2030. Interestingly, Asia is where almost half of urban population of the world lives and soon it will have the majority of the world's urban population.

It is now well established that the acceleration of urbanisation generally takes place with corresponding acceleration of economic growth. Urbanisation is promoted by (i) economies of scale in production particularly in manufacturing; (ii) the existence of information externalities; (iii) technology development, particularly in building and transportation technology; (iv) substitution of capital for land as made possible by technological developments. Information asymmetries contribute to agglomeration economies. As economies of scale in production begin to take hold larger size plants become necessary, thus contributing to the need for larger settlements of people. The services needed by the rising agglomeration of people gives rise to an even greater number of people living together: thus cities are born and how they grow. As technology develops and capital is substituted for land, taller buildings become possible, intensifying population densities further. Similarly, technology development in transportation, enabling faster speeds, enables people to live at greater distances thus contributing to the expansion of city size. The existence of agglomeration economies gives rise to continuing accretion of people in a settlement, thus promoting city growth. These linkages become more prominent with economic growth thereby promoting the acceleration of urbanisation. With the growing weights of industry and services in developing countries, urbanisation has proceeded apace over the last 50 years. As we will see, the relatively concentrated pattern of Asian urbanisation that has occurred along with a very high rate of economic growth perhaps best illustrates the economic gains accruing from agglomeration economies and economies of scale.

Table 2.3 Global Urbanisation Trends—Level of Urbanisation

		(Per cent)
Region	**1920**	**2030**
World Total	19	61
Less Developed Regions	10	57
Africa	7	54
Asia	9	55
Latin America	22	85
More Developed Regions	40	85
Europe	46	83
North America	52	85
Oceania	47	75

Source: Mohan and Dasgupta (2005).

ASIA BECOMES URBAN

Presently the highest rates of economic growth are being witnessed in Asia, and hence high urban growth. This is particularly noticeable in China and India, which today have the largest rural populations but are urbanising rapidly. Even in other Asian countries where a large number of cities are witnessing high rates of economic growth, the growth in their urban population is also going to be higher. Of the 10 most populous countries, 6 are in Asia (Table 2.4). Even as more than two billion people will be added to Asia's population in the next 30 years many of these countries will still be rapidly urbanising. This phenomenon has now gained such prominence that even popular news magazines have begun to take note (Newsweek, 2003). This is also reflected in Figure 2.1 which shows that rural population in Asia is expected to decline, in absolute terms during 2000–2030, yet another unprecedented event, while urban population will almost double during the same period. Therefore, while thinking and thought processes related to urbanisation were dominated by the growth of urbanisation in Latin America between 1950 and 2000, the 21st Century will be the "Asian Urban Century".

By 2030, Asia alone will have about 2.7 billion urban people accounting for over 50 per cent of its total population. All the other regions of the world will have a combined urban population of about 2.3 billion. Six of the 10 countries with the largest urban populations are in Asia. Urbanisation in Bangladesh has been among the fastest in the world (5.6 per cent during 1950–2000). The only

Table 2.4 Emergence of Mega Cities: Population of Cities with 10 Million or More Inhabitants

(in Millions)

1950		1975		2000		2015	
New York	12.3	1 Tokyo	19.8	1 Tokyo	25.2	1 Tokyo	27.2
		2 New York	15.9	2 Sao Paulo	18.3	2 Dhaka	22.8
		3 Shanghai	11.4	3 Mexico City	18.3	3 Mumbai	22.6
		4 Mexico	10.7	4 New York	16.8	4 Sao Paolo	21.2
		5 Sao Paulo	10.3	5 Mumbai	16.5	5 Delhi	20.9
				6 Los Angeles	13.3	6 Mexico City	20.4
				7 Kolkata	13.3	7 New York	17.9
				8 Dhaka	13.2	8 Jakarta	17.3
				9 Delhi	13.0	9 Kolkata	16.7
				10 Shanghai	12.8	10 Karachi	16.2
				11 Buenos Aires	12.1	11 Lagos	16.0
				12 Jakarta	11.4	12 Los Angeles	14.5
				13 Osaka	11.0	13 Shanghai	13.6
				14 Beijing	10.8	14 Buenos Aires	13.2
				15 Rio de Janerio	10.8	15 Metro Manila	12.6
				16 Karachi	10.4	16 Beijing	11.7
				17 Metro Manila	10.1	17 Rio de Janerio	11.5
						18 Cairo	11.5
						19 Istanlbul	11.4
						20 Osaka	11.0
						21 Tianjin	10.3

Source: United Nations (2002).

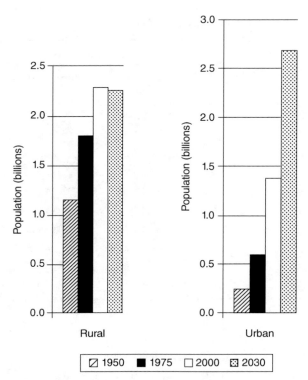

Figure 2.1   Growth in Urban and Rural Population in Asia Over the Years.
Source: UN, 2002

African country in the top 10 list, Nigeria, has also urbanized very rapidly in the past 5 decades (World Urbanisation Prospects, 2001). Interestingly, by 2010, Lagos is projected to become the third largest city in the world.

Moving to the city level urbanisation trends, the growth of urban agglomerations in developing countries has far exceeded that in developed countries. In 1950, there was only one city with a population of over 10 million people: New York City. In 2000 there were 17 cities with a population above 10 million, 22 cities with population between 5 and 10 million; 402 with a population of 1 to 5 million; and 433 in the 0.5 to 1 million category. An important characteristic of urbanisation in Asia has been the emergence of mega-cities—large multi-nuclear urban agglomerations of more than 10 million people. There were no such agglomerations in Asia in 1950, two in 1975 and by 2000, 10 of the 17 global megacities were in Asia. It is expected that 12 out of the 21 mega cities in the year 2015 will be in Asia (Table 2.4).

Of the 21 cities expected to reach 10 million plus by 2015, 17 of them will be in developing countries. Of these 17 cities, 11 of them will be in developing countries in Asia and those will be Dhaka, Mumbai, Delhi, Jakarta, Kolkata, Karachi, Shanghai, Metro Manila, Beijing, Istanbul and Tianjin (Table 2.5).

The historical pattern of urbanisation suggests that countries tend to urbanise very slowly until they attain urbanisation levels of around 25 to 30 per cent. The pace of economic growth and overall development then quickens, with rapid structural shifts occurring in the economy, away from agriculture to industry and services. The pace of urban growth between urbanisation levels of 25 to 30 per cent and 55 to 60 per cent has typically been observed to take place in a very short historical time frame of 25 to 50 years. This happened in European countries and North America at different times in the late nineteenth century and early twentieth century, and in Latin America during the latter half of the twentieth century. Japan went from about 25 per cent urbanisation level in 1930 to over 70 per cent in 1970; and Korea from about 25 per cent urban in 1955 to 50 per cent by 1975. During this rapid phase of urbanisation, the demand for urban infrastructure investment is massive and countries have usually been observed to need external savings to supplement available domestic

Table 2.5   Out of the 10 most Populous Countries 6 are Asian

		1950		2000		2030	
	Country	Per cent urban	Population (mn)	Per cent urban	Population (mn)	Per cent urban	Population (mn)
1	China	12.5	555	35.8	1275	59.5	1485
2	India	17.3	357	29.0	1009	40.9	1409
3	USA	64.2	158	77.2	283	84.5	358
4	Brazil	36.5	54	81.2	170	90.5	226
5	Indonesia	12.4	79	41.0	212	63.7	283
6	Nigeria	10.1	30	44.1	114	63.6	220
7	Pakistan	17.5	40	33.1	141	48.9	273
8	Mexico	42.7	28	74.4	99	81.9	135
9	Japan	50.3	84	78.8	127	84.8	121
10	Bangladesh	4.3	42	25.0	137	44.3	223

Source: United Nations (2002).

resources to finance such investment. So far, the world has been able to cope with such demands as the scene of urbanisation has shifted from one region to another and the overall magnitude of increase in urban population has been manageable. What is new in the next 25 to 30 years is that three of the world's most populous countries, China, India and Indonesia, with combined total population of about 2.5 billion will be undergoing this process simultaneously, with Pakistan and Bangladesh not much far behind. The magnitude of the increase in urban population in Asia in this period will be unprecedented and will undoubtedly give rise to unforeseen problems as well as opportunities.

The popular view of towns and cities in developing countries, and of the urbanisation process, is a negative one, despite the benefits it brings. For many, the emergence of such cities connotes environmental degradation, the generation of slums, rampant urban poverty and unemployment, loss of control and traffic chaos. But what is the reality? Given the unprecedental increase in urban population over the last 50 years from 300 million in 1950 to 2 billion in 2000 in developing countries, the wonder really is how well the world has coped, and not how badly. In general, the urban quality of life has improved in terms of availability of water and sanitation, power, health, education, telephones, and the like, and poverty has fallen. These improvements must be viewed against the fact that they have been achieved in the presence of rapidly increasing population, under difficult fiscal situations, and with strained human resources for the emerging needs of public management.

By way of illustration, we can look at the coverage of water and sanitation services in Asian cities. As presented in table 2.6 a large number of urban residents have been provided with improved water in urban areas in Asia's largest countries. Although in some countries such as China, Indonesia and the Philippines, the access to improved water in terms of percentage of total urban population seems to have declined from 1990 to 2000, in absolute numbers, millions of additional citizens have actually been provided improved services. Although the definition of urban areas as well as the concept of improved water services varies across the different countries referred to in the following tables, the increase in access is documented by each country within their own definitions as presented here. In the four countries taken together, an additional population of approximately 262 million has been provided with improved access to water in urban areas during the 1990s, which is a population greater than that of most countries today.

These countries in Asia have made significant progress in the provision of sanitation services too, together providing for an additional population of more than 293 million citizens within a decade. In this parameter there has been a consistent increase in the percentage of urban population covered in all the countries (Table 2.7).

Table 2.6 Improvement in Access to Water in Urban Asia

Country	Per cent of urban population with access to Improved water		Additional people provided (million)
	1990	2000	1990 to 2000
China	99	94	115
India	88	95	107
Indonesia	92	90	27
Philippines	93	91	12
Korea, Rep.		97	

Source: World Bank database, 2003.

Table 2.7 Improvement in Sanitation Facilities in Urban Asia

Country	Per cent of urban population with access to Improved sanitation facilities		Additional people provided (million)
	1990	2000	1990 to 2000
China	57	68	130
India	44	61	96
Indonesia	66	69	23
Philippines	85	93	15
Korea, Rep.		76	29

Source: World Bank database, 2003.

This general improvement in access to urban infrastructure and services in the Asian cities can be observed in other sectors too such as health services, education and housing. Looking at the change in poverty levels too is revealing; in terms of both nutrition levels and life expectancy most of the Asian urban areas have seen consistent progress. However, in income poverty terms the experience has been mixed, bringing to the fore the importance of macro-economic management of a country, and its relevance for urban poverty reduction. In India, a steady macro economic environment and economic growth in the 1990s has led to steady improvement in income poverty levels.

Progress in the provision of essential urban services has been significant. Unlike the popular view that urbanisation causes deprivation, urbanisation has been relatively well addressed in Asian cities and has led to an increase in living standards and quality of life of its residents. Given the (i) vast increase in urban population in these countries, (ii) low per capita income, (iii) constrained fiscal circumstances of governments, leading to low expenditure on urban infrastructure, and (iv) the existence of weak local governments in most urban areas, the progress achieved is indeed quite noteworthy. In all probability the quality of life in developing Asian urban areas is significantly better than the situation witnessed in the 18th and 19th centuries in European cities, which had grown under similar circumstances, but perhaps at higher prevailing income levels. We may also recall that they did not have to cope with mega cities during their phase of rapid urbanisation. These achievements have probably been enabled by the availability of better technology and systems now.

SOME DISTINCTIVE FEATURES OF ASIAN URBAN GROWTH

The rapid economic growth of Asia in the last half century must be amongst the most spectacular periods of development in recorded human history. The magnitude of population that has benefitted from this growth far surpasses that of the rest of the world, and in particular of Western Europe and North America. Broadly speaking, the evolving pattern of Asian urbanisation has naturally corresponded to the shifting focus of economic growth over this period.

Economic growth in Asia was kindled by the remarkable post World War II recovery of Japan in the 1950s and 1960s, and stretching into the 1980s when Japan became the second largest economy in the world. A particular characteristic of Japanese economic and urban growth was the heavy concentration of economic activity in the Tokkaido region (Tokyo—Nagoya—Osaka corridor), which was aided and abetted by an apparent conscious choice of concentrated infrastructure investment in this region. The Japanese economy benefited from high savings and investment rates (almost 40 per cent of GDP by 1970) during this period, which provided the resources necessary for the heavy transportation and urban infrastructure investments that were made. The rapid increase in manufacturing investment and production gave rise to high growth in manufacturing employment that was responsible for very high rates of rural urban migration. The Japanese countryside was literally drained of people during this period and the Tokkaido region became one of the most densely populated urban corridors in the world. Between 1950 and 1970 Japan's rural population fell from its peak of about 52 million to less than 30 million; by then almost 40 per cent of Japan's total population, and as much as 60 per cent of its urban population was concentrated in the 500 km Tokkaido coastal corridor (Mills and Ohta, 1970). The kind of economic concentration that emerged was perhaps instrumental in economizing on infrastructure investment that would have been larger had it been spread out over a larger part of the country. The geographical proximity of different activities gave rise to agglomeration economies that aided rapid productivity growth and also enabled innovation in traditional production processes through the introduction of new systems such as Just-in-Time (JiT) modes of inventory management. Such innovations enabled much more outsourcing of components, a process that contributed to the drastic reduction in manufacturing costs that was the foundation of Japan's competitiveness. The more efficient inventory management resulting from JiT and overall supply management has also enabled significant reduction in corporate need for bank financing, leading to significant changes in bank portfolios. Creativity and innovation have been among the distinctive characteristics of Japanese economic and urban growth.

The focus of growth began to shift later to the Asian tigers: Hong Kong, South Korea, Singapore and Taiwan. What is noteworthy is that the overall pattern of growth was similar in these countries. Singapore and Hong Kong being city states were, of course, somewhat different and had to exhibit concentrated growth. But South Korea and Taiwan also concentrated on specialising in manufacturing. Like Japan they first specialized in labour intensive low technology goods production and then began to move up the technology chain. While Korean manufacturing production was concentrated in large manufacturing conglomerates, that of Taiwan was spread over a large proportion of small and medium enterprises. However, they both exhibited a strategy of concentrated spatial development in urban concentrations, Seoul/Pusan in Korea and Taipei/Kaohsing in Taiwan. The Seoul and Pusan metropolitan regions accounted for almost 70 per cent of South Korea's urban population by the mid 1970s (Mills and Song, 1979 p.188). Each of the Tigers adopted an export-oriented and outward orientated strategy, which also necessitated heavy investments in key transportation and communication links with the rest of the world. The economic activities located in these cities were as connected with the rest of the world as with their hinterlands, if not more. This common heavy spatial concentration in these countries, Japan, South Korea and Taiwan, can perhaps also be attributed to the fact that they are among the most densely populated countries in the world.

Given the success of the tigers, it was then the turn of the cubs, the South East Asian countries of Thailand, Indonesia and Malaysia during the 1980s. Once again, the pattern of concentrated heavy investment was repeated in the metropolitan cities of Bangkok, Jakarta and Kuala Lumpur. The orientation here was also export oriented and hence, again, heavy investment had to be incurred in transportation and communication links, and in urban infrastructure.

The urban development pattern that emerged in Asia was that of concentrated development around coastal regions of each country. Moreover, the transportation links that grew between these coastal cities in terms of economic linkages of communication, transportation and commerce, contributed to the emergence of trans-border virtual urban corridors. In fact, a look at the Asian urbanisation pattern as it has emerged reveals a long almost continuous urban coastal corridor stretching from Tokyo to Sydney, through Seoul, Taipei, Shanghai, Hong Kong, Kuala Lumpur, Singapore and Jakarta (Douglas, 1998).

Interestingly, early Chinese economic and urban growth in the 1980s and 1990s was also the result of a similar strategy: export-oriented labour-using manufacturing located in the coastal areas; initially in and around Shanghai and in the whole Pearl River Delta Region. Once again, infrastructure investment was concentrated in the Special Economic Zones and, as in the other countries, heavy rural urban migration ensued. Given the size of China it is, of course, not easy to portray its pattern of urbanisation. Until the 1990s, rural urban migration was heavily constrained through the urban residence permit system, which has since been loosened considerably. Although there is a great degree of debate on the actual level of Chinese urbanisation, in 2000 it was somewhere between 30 and 36 per cent. It now has about 90 cities with more than 1 million population. The Chinese authorities have a clear priority objective of accelerating urbanisation to absorb surplus labour from rural areas into more productive urban systems (Webster, 2005).

In contrast to this common experience, the Indian strategy was almost a mirror image of the East and South East Asian strategy. The ethos was of dispersed development: urban concentration was frowned upon and actively discouraged; and the import substituting inward oriented manufacturing approach persisted till the 1980s. Investment in infrastructure, particularly urban infrastructure has been of lower intensity. As I have documented elsewhere, unlike the East and South East Asian countries, during the period of accelerating economic growth in India in the 1980s and 1990s, although industrial growth was high, manufacturing employment and urban population growth decelerated (Mohan and Dasgupta, 2004). Despite being a peninsula with a long coastline, there was no attempt to concentrate economic activity in the coastal areas: in fact growth in the old concentrations of Calcutta, Madras and Bombay (now Kolkata, Chennai and Mumbai) slowed down in the 1980s and 1990s while, interestingly, inland cities such as Bangalore, Hyderabad and Delhi prospered. There have perhaps been few such examples of inland cities growing faster than coastal cities and regions.

The fulcrum of global economic growth has now shifted to the large economies of China and India. In the case of China, with the initial growth impetus having come from the coastal zones, the emphasis is now shifting to inland cities. Although economic growth has perhaps now got more concentrated regionally in India, there is still little evidence of strategy shifting to the promotion of greater urban concentrations. So the export led, coastal urban growth that has been characteristic of Asian urban growth in the last 50 years can now be expected to move inland in the large land masses of India and China. Such a pattern of urban growth will probably necessitate higher degree of infrastructure investment—both intra-urban and inter-urban in order to ensure international economic competitiveness.

EMERGING ISSUES FOR THE NEXT WAVE OF ASIAN URBANISATION

By all accounts Asia has coped well with the unprecedented magnitude of urbanisation that it has experienced in the last 50 years. The Asian habitat pattern has been transformed over a historically brief time period: an Asian is now almost as likely to be found living in an urban area as in a rural area, with a high probability of being found in a city of significant size. Because of the particular economic strategy followed over much of Asia, its cities are engines of economic activity exhibiting ever increasing productivity gains and prosperity. This has enabled Asia to finance its urban infrastructure investment without excessive international borrowing. In fact, the financial surplus that the region is now exporting to other regions of the world has come as a bit of a surprise, given its own resource needs for continuing investment, particularly in the infrastructure needed for further urbanisation.

Although the rate of urbanisation will, no doubt, slow down overall, the magnitude of urban population accretion in Asia over the next 30 years will be roughly equal to that experienced in the last 50 years. In fact, this next wave of urbanisation in Asia will be the largest in magnitude over any 30 year period in human history. The key question that arises is whether the region will have enough resources to cope with this magnitude of urbanisation. It is the most populous countries of China, India and Indonesia, along with Pakistan and Bangladesh, which will undergo large urbanisation during this period, even though the pace of change may well be faster in other countries such as Vietnam, Laos and Myanmar.

China seems to have invested adequately in infrastructure already and there would appear to be little doubt of its ability to generate enough internal resources to finance its investment needs over the foreseeable future and its ability to attract external resources if needed. In fact, given the current magnitude of its current account surplus, coupled with the flow of external savings into the country, and the large magnitude of forex reserves invested elsewhere, it has enough of a cushion to meet most if not all its needs in the foreseeable future. As I have discussed, the change that can be expected is that of a change in focus towards inland cities. The question that then arises is whether these cities will be productive and competitive enough to produce the economic surpluses necessary for their continued sustainability. The attainment of such productivity will necessarily mean greater inter-urban infrastructure investment, so that these cities are well integrated with their coastal cousins. Furthermore, given the information technology and communication revolution, along with the secular decline in per unit air transportation costs, they can now also be connected to the rest of the world without intermediation of the coastal cities. But this would also mean that they will need to specialize more in service industries, rather than in manufacturing; the latter could be handicapped in global competitiveness because of excessive transport costs. Thus, it is certainly the case that much greater inter-urban infrastructure investment will be necessary to make these cities competitive. It would appear that this has already begun in China in all the various facets of infrastructure: road, rail, airports and telecommunications.

The Indian story is somewhat different with relatively low attention being paid to urban development over the years, and slowing urbanisation over the last quarter century. There have been systematic policy biases against labour using industrialization, location of industries in urban areas, and against urban concentrations. Correspondingly, India has severe problems in both the management and financing of cities. With the new found economic resurgence of India, a result of consistent economic reforms since the early 1990s, the importance of urban infrastructure investment has finally begun to occupy the minds of key policy markers and a new "National Urban Renewal Mission" has been launched. However, the biases against labour using manufacturing remain in overall economic policy making, in the industrial regulatory regime, labour regulations and in urban land policy. Thus employment growth in manufacturing remains low. As I have argued elsewhere, these polices could have contributed significantly to the slow down observed in Indian urbanisation over the last quarter century (Mohan and Dasgupta, 2004).

Industrial competitiveness in India has now recovered after the shock of competition having been absorbed through significant financial and business process restructuring in Indian firms. The export orientation of Indian industry at large has also increased significantly in recent years. On average (since 2000–01), about 14 per cent of sales of Indian firms is now exported and this proportion continues to grow, as compared with 7 per cent in 1991–92. Consequently, the continued high growth of the Indian economy would get strengthened if the efficiency of Indian cities improves. It is noteworthy that those cities that have shown great economic vigor over the last decade in India, such as Delhi, Bangalore, Hyderabad, Pune and Chandigarh exhibit certain common characteristics. They have an unusually large endowment of educational institutions at all levels, and research institutions. As it happens, a good number of relatively high technology public sector industries were also located in most of these cities. With the availability of such an educational and technical ethos, these cities have a knowledge base that is significantly superior to that of other cities. Consequently, they have been able to lead the Indian information technology revolution and to benefit from all the high economic growth that has followed. The lack of appropriate physical infrastructure and transport linkage inland or with rest of the world has not come in the way since the information technology exports are not dependent on these elements of infrastructure. All they needed was appropriate communication infrastructure, which has indeed been provided progressively. However, the prosperity brought by the success of the IT industry in these cities has itself resulted in greater pressures being placed on the existing infrastructure. Traffic congestion has arisen due to much elevated levels of auto ownership; housing demand has escalated in both quantity and quality leading to rapid increases in land and housing prices; and much increased power demand is putting great stress on existing power supply systems. Businesses are therefore beginning to look for other locations. The competitiveness of these cities will therefore depend on acceleration in urban infrastructure investment and improvement in urban governance and management. The successful financing of such an enhanced level of investment will depend crucially on the financial viability of such projects.

Given that the level of Indian urbanisation is still less than 30 per cent, and that 60 per cent of the Indian total population is still dependent on agriculture, the continuation of high economic growth will depend on the success of a higher rate of labour absorption by cities. This will need much higher growth in labour using manufacturing, higher levels of urban infrastructure investment, along with knowledge based forward looking city management. Thus, the Indian situation is quite different from that of China. If Indian urbanisation does speed up as it surpasses the 30 per cent mark and as Indian per capita income approaches US $1000, in normal circumstances we should expect acceleration in Indian urban growth. This will need significant acceleration in urban infrastructure investment and hence in the mobilisation of financial resources for such investment. Although so far India has not relied significantly on external savings for its investment needs, it is possible that the demand for urban infrastructure investment will necessitate greater usage of external savings during this phase of India's urban growth.

The other large country in Asia is Indonesia, which is spread over a large number of islands. Till the Asian financial crisis in 1997, Indonesia also exhibited economic policy characteristics similar to those of other East-Asian countries in terms of openness and export orientation, although it had more of a mix in economic policy, which was also concerned with promotion of import substituting industries and conscious dispersal of economic activities beyond the natural concentration in Java. Nevertheless, the greater Jakarta region, known as the Jabotobek region, exhibited a high degree of urban concentration where a high proportion of Indonesian economic activities got concentrated despite the large size and dispersed nature of the Indonesian archipelago. Given the somewhat lower level of Indonesian per capita income and the very rapid growth of the Jabotobek region, the extent and proliferation of slums has been high in the region. (Webster, 2004) Furthermore, Indonesia was perhaps the most highly affected of the Asian countries from the 1997 East Asian financial crisis. It has still to fully recover from that shock and is yet to regain earlier economic dynamism. Thus the persistence of slums and the accompanying urban distress is likely to persist in Indonesia longer than its other South East Asian counterparts. The urban future of Indonesia is more beset with uncertainties, reflecting the parallel economic uncertainties that it faces.

How do we then look at Asia's evolving urban future over the next 30–50 years? How will it be different from the experience of the last half century? The one key difference is that with increasing globalisation and ever higher levels of income that the region as a whole is now blessed with, the residents of Asian cities will now be much more demanding relative to their predecessors in terms of the quality of urban services that they deem to be their right and the urban amenities of living that are now seen as normal. Hence it is likely that urban investment will be different in terms of its composition and intensity. Second, with increasing globalisation and reduction in trade protection, each Asian city will have to be more competitive on a global scale than has been the case in the past. In the larger countries there will be inevitable tension between the claims of coastal urban areas that possess natural comparative advantage and the vast hinterland that will need greater infrastructure investment for attaining competitiveness. Hence, policy makers probably need to give greater explicit attention to the ingredients of competitiveness, the corresponding public investment that will be appropriate in this regard, and the modes of financing that will need to be mobilised. Third, as discussed, Asian urbanisation in the last half century has been based disproportionately on rapid city-based manufacturing growth in labour intensive industries that have pulled in labour from rural areas, thereby relieving rural areas of excess labour and hence enabling growth in both rural and urban productivity. With the changes in technology that have now taken place it is an open question whether labour intensive industry will continue to survive and grow in the manner and experience of the previous 50 years, and whether it will be as easy as in the past for urban areas to absorb the kind of rural urban migration experienced earlier.

This issue is of great importance to India since the share of manufacturing in its economy is somewhat lower than could be expected at its current level of economic development (Mohan, 2002). If India is not able to change its economic and urban specific policies to encourage labour using manufacturing in and around urban concentrations, and if the global economic imperative is that such patterns of industrialization are no longer feasible, how will its cities grow and absorb the large rural population that needs to get off the farm so that both rural and urban productivity can grow faster? Thus, we can expect the pattern of Indian industrialization and urbanisation to be different from that of East and South East Asian countries.

It must still be understood that for successful and sustainable urbanisation the share of manufacturing will still need to increase, but with somewhat different characteristics (Yusuf and Nabeshima, 2006). The manufacturing process has itself changed significantly so that many activities that were earlier concentrated in one location in one plant are now often outsourced to many different locations within an urban concentration and even across borders. Often product design is now increasingly information technology dependent and typically locationally divorced from the core manufacturing plant. Moreover, product development and design is now increasingly being outsourced on a global basis. The availability of competent engineering skills at a lower cost in India is contributing significantly to the relocation of product development and design from developed countries to India (Marsh, 2006, a,b,c). Conversely, Indian manufacturers are also outsourcing their product development and design in the reverse direction. Second, increased global competition is also leading firms to look for continuous reduction in core manufacturing costs in whatever ways that are practical. Local outsourcing of components and processes is one of the common practices that has been found to be useful in this regard. The requirements of inventory control and management require the location of such outsourced manufacturing activities to be in close proximity to the mother plants. Thus, successful industralisation in this manner in India would increasingly require greater concentration of these activities than has been experienced in the past. With the quality of manufactured goods improving all the time, it is becoming clear that the demand for low skilled labour is unlikely to accelerate. Hence, a core component of economic and urban policy would have to be enhancement of skills of the labour force at all levels. The provision of vocational training has been difficult in most countries since successful training needs to be market determined; but the private sector often finds it difficult to design an appropriate revenue model and public provision is typically not market sensitive. The need is for public private partnership, which is not easy to design. Successful urbanisation in the future will be crucially dependent on the availability of labour with appropriate skills.

Thus for Asian cities growing in the next 30 to 50 years it is becoming increasingly clear that the key to their success will indeed lie in the continuous enhancement of human resources. In the globalizing world, creativity and entrepreneurial dynamism will be the essence of successful cities (Yusuf and Nabeshima, 2006). All the East-Asian cities, such as, Bangkok, Beijing, Singapore, China, Seoul, Tokyo and others exhibit high levels of educational attainment, and have impressive endowments of educational and research institutions. In fact, it is noteworthy that some of these cities, such as Hong Kong and Singapore, which did not traditionally have higher education institutions that were particularly noted for high quality, have in the last two decades consciously invested intensively in higher education institutions in terms of both quantity and quality. Each of these major Asian cities now houses large numbers of universities. Illustratively, Tokyo has 113 universities and Beijing has 59 although, there is a great deal of variation in the quality of these universities (Yusuf and Nabeshima, 2006). Similarly, in India, it is the southern region where a large number of private colleges and universities have emerged to cater to the increasing demand from industry for technical personnel. Thus, apart from the traditional needs for physical urban infrastructure investment for successful urbanisation, similar attention now has to be given to the soft infrastructure that is related to the creation, production, and retention of knowledge, along with facilities that enable continuous skill enhancement.

Openness to the outside world does not just mean increase in trade in goods and services. It also means greater openness to ideas and new practices. In a recent conference on "Urban Dynamics in New York City" organized by the Federal Reserve Bank of New York (Yes, Central Banks are interested in city growth), Kenneth Jackson attributed the great success of New York city to its openness to new waves of immigrants over time. "The constant infusion of new energy and ideas into the metropolis over the years enabled New York to meet economic and technological challenges that destroyed the prospects of competing cities" (Jackson, 2005). It is quite remarkable that most of the successful East and South-East Asian cities have remained very open to the presence of foreign citizens with high levels of education and skills. There are said to be almost 100,000 foreign citizens in Beijing alone (Yusuf and Nabeshima, 2006). Such presence of foreigners contributes greatly to the economic vitality so needed by growing cities, as it provides new competition to residents, while facilitating the flow of new ideas in both directions. In fact, a large number of universities and other technical institutions in the developed world have also begun to realize that it would be increasingly efficient for them to relocate some of their activities to Asian cities rather than drawing Asian personnel to their parent campuses. Thus, enhancement of human capital at different levels will evolve different strategies and increasingly greater openness to cross-border flow of institutions and personnel.

THE CHALLENGES OF URBANISATION IN THE TWENTY FIRST CENTURY

Of the total projected increment to world urban population between 2000 and 2030 of about 2.1 billion people about 1.3 billion or about 60 per cent, will be in Asia (Table 1). In the second half of the twentieth century the total accretion to urban population in the world was similar in magnitude (about 2.1 billion), but the Asian share was somewhat lower at about 53 per cent. As I have repeatedly emphasized, it is this expected magnitude of urbanisation expected in Asia that is unprecedented and hence the management of it is the key challenge facing us in all its multifaceted aspects.

I have attempted to speculate about the possible changes in urbanisation patterns and requirements that may emerge in the future. Just as the structure of American cities is different from that of European cities depending on their vintage, we can expect the 21st century Asian city to also exhibit different characteristics. The older European or Asian city is typically more densely populated and less spread out than the American cities, reflecting in particular the different degrees of motorization that existed at their inception. American cities are much more dependent on privately owned motorized transportation than cities in the older continents. Even as early as the early 1970s, nearly 80 per cent of US urban commuters traveled by car, as compared with only 15 per cent in Japan. In fact, 65–70 per cent of Tokyo commuters and 60 per cent of those in Seoul traveled by public transit in the early 1970s (Mills and Ohta, 1976; Mills and Song, 1979). With increasing incomes and aspirations the pace of growth in auto ownership in Asian cities is awesome as is the growth in traffic congestion.

The current increase in oil prices is sharpening the kind of tensions that are typical in debates related to urban transportation. With the emergence of increased private motorized transportation there has been a noticeable intensification of investment in intra-urban expressways in many Asian cities. This typically leads to accelerated urban sprawl, a still faster increase in auto ownership and consequent demand for oil, and higher pollution. Over time, road traffic congestion inevitably catches up leading to further demand for road investment or for mass transportation that is then expected to reduce road traffic congestion and pollution. The current trends suggest that the result is high investment in both modes of transportation. Given the higher levels of income that already exist, and expected rapid increase in income growth, the emergence of these patterns is perhaps inevitable. The demand for both financial and physical resources will clearly intensify, and the question is whether it will be possible for Asian cities to impose appropriate taxation systems and user charges that can finance the investments required.

It is widely accepted that the current surge in oil prices is more demand related than to disruption in supply as was the case earlier. With the expected pace of Asian urban growth over the next 30 years do we then expect ever increasing oil price increases in response to ever increasing demand? Or will there be corresponding supply response, as has been the case in the past, which will contribute to oil prices falling again? In either case, appropriate petrol pricing and urban transport pricing will be as crucial for urban policy as for economic policy as a whole. As is well known the emerging transportation pattern also affects city structure crucially: so urban transport policy will be of great importance to the kind of growth that we observe in Asian cities in the coming years. Will the growing Asian cities be a new amalgam of the typical old densely populated city centre, co-existing with suburban sprawl characterised by motorized transportation modes, shopping malls akin to the American pattern. In some Asian cities, the old city centres are begin completely reconstructed as in Beijing and Kuala Lumpur, whereas in others the tension between the old and new continues.

Another general issue affecting the pattern of urbanisation will be the nature and pace of rural urban migration. In the case of China, because of the long standing one child policy, natural growth of urban population is low and hence the same rate of urban population growth gives rise to a much higher order of rural urban migration than in other countries such as India where the natural rate of urban population growth is higher. In the former, household size is presumably smaller and hence, for the same population size, investment in housing and associated infrastructure will have to be higher per capita. Furthermore, the cultural problems associated with first generation migrants are likely to be greater. Conversely, it is also possible that with higher natural urban population growth there could be greater local resistance to in-migrants giving rise to associated problems in economic and social policy. Thus policy makers also need to give attention to the specific nature of economic demographics in their respective countries as it affects urbanisation.

As the South East Asian countries urbanized fast in the 1970s and 1980s the problems of slums and associated deficiencies in urban infrastructure services related to water, sanitation, sewerage and solid waste disposal received great attention. Because of the high economic growth these problems have taken care of themselves in many countries. However, in still low income populous countries such as India, Bangladesh, Pakistan and Indonesia the existence of slums and lack of services remain a serious issue. Given the large numbers of people involved, issues related to change management are as important as those related to financing and resource management. Flexibility in urban land policy and zoning, the working of land markets, availability of housing finance, and facilitation of urban land development all need focused attention. The availability of sympathetic policy makers and professionals in these areas is at a premium in these countries. Generating skills and professionalism in urban management in all its aspects will therefore be among the key challenges that Asian urbanisation will bring in the coming years.

One of the consequences of globalisation, more open trade in both goods and services, and the vastly increased trans-border mobility of the professional classes, has led to the prevalence of international compensation levels for these groups despite lower average income levels in Asian cities, and hence to greater inequality in these cities. Members of these "creative classes" (Yusuf and Nabeshima, 2006) also seek an assortment of urban attributes that were not demanded earlier. They are much more demanding in terms of quality of housing and urban services, health and education services. Knowledge workers are also very interested in the availability of recreational amenities, cleanliness of environment, efficient and comfortable transportation, and international level communication services. Thus, in order to attract and retain the very people who are essential for city competitiveness, Asian cities will have to prematurely invest in world class facilities at much lower average income levels. The most competitive of Asian cities have clearly recognized this as is evident in cities such as Shanghai, Hong Kong, Singapore, Kuala Lumpur and Seoul, with Bangkok fast attempting to catch up.

The task for policy makers for managing Asian urbanisation in the next thirty years is therefore more complex than in the previous fifty. In addition to the traditional problems of providing, financing and managing basic physical infrastructure, they have to be more conscious of the emerging demands resulting from globalisation in terms of creating knowledge based cities that also boast of competitive urban amenities. Increase in the sheer number of large cities over any cut off point, one million, five million, or ten million, will also stretch the capability of government authorities of finding people with appropriate skills for city management. Here, it is perhaps correct to say that there has been some decline in international attention to the generation of such skills, and may well be an area for coordinated international attention.

Finally, being a central banker, I can hardly conclude this address without making some remarks on the financing needs of Asian urbanisation over the next thirty years. Urban infrastructure typically lasts for long periods of time. Hence, while urban infrastructure investment has to be made ex-ante at the time of rapid urban growth over a period of 10–15 years, benefits may well flow for periods as long as fifty years or more. Life would be easy if financing sources were such that civic authorities could raise resources in such a manner that the repayment schedule matched the benefit schedule. A scan of urban financing systems across the world does not reveal any uniformity in pattern. Germany has used its mortgage banks to sell Pfandbrief bonds that enjoy high credit quality next only to the Bond, and then intermediate the funds to states and municipal authorities for infrastructure investment. There is a complex system of credit enhancements that makes it feasible to raise long term funds. But this credit quality has been earned over more than a century over which the municipal authorities have made sure that their tax and user charges systems are such that they can redeem the resources raised. In the United States, it is the decentralised municipal bond system that has largely financed urban infrastructure. Here also, since the ability to raise resources depends on the retention of healthy credit ratings, municipal authorities have a very strong incentive to stay solvent and service their bond holders. In principle, therefore, such systems have been successful since they have ensured that towns and cities face on incentive structure that encourages them to remain creditworthy and are essentially self-financing.

In Asian countries, financial markets have not been sophisticated enough to allow for such financing methods yet. Financing for urban infrastructure has usually come for higher tier governments who raise resources from taxes, or from banks and financial institutions that have been typically government owned or sponsored. Such systems are not well designed to avoid moral hazard: the recipient towns and cities do not have as strong an incentive to be essentially self financing. The 1990s have seen increasing attempts to privative the provision of urban infrastructure, but this has met limited success at best. Given the magnitude of urban population accretion expected over the next thirty years, I see little choice. If Asian cities are to thrive and prosper, they will have to develop self sustaining local taxation and user charge systems so that they can tap national and international financial markets for their financing needs.

This brings me to the international dimension of urban infrastructure financing. It is usually the case that, when a country begins its rapid urban growth phase its financial markets are yet to develop, the only way to tap long term funds is to take recourse to external savings, which are then to be repaid over a long period of time. The typical historical experience has been that the regions undergoing intensive urbanisation had to mobilise external savings intensively; followed by periods of balance of payments crises and debt defaults. In Asia, too, the 1997 financial crisis was also partially reflective of large external resource flows earlier that suddenly got reversed, as was the Latin American debt crisis of the 1980s. Since then, however, it is puzzling that the region as a whole is exhibiting financial surpluses that are being invested in Europe and North America. In the great current debate on global imbalances, the assumption seems to be that these imbalances seem to be of a relative durable nature, partly reflecting the favourable economic demographics of Asia and the converse in the West.

I remain somewhat puzzled by this financial turn of events. I would have expected that, the demands of infrastructure investment, particularly that of urban infrastructure, would be such that regional domestic savings will not be adequate to finance the required investment. Perhaps the explanation really lies in the Asian reaction to the 1977 financial crisis and that we may expect higher investment levels in the years to come. The magnitude of urban population growth expected in China, India, Indonesia, Pakistan and Bangladesh expected over the next 30 years is such that pressures on international resource mobilization are bound to arise. Urban infrastructure investment would then exceed available savings in these countries and the current alleged savings glut will disappear over a period of time. Will there then be enhanced competition among Asian countries from available international savings? With the emerging adverse demographics in the West, and hence low savings rates there, will this competition lead to the emergence of higher real interest rates in the years to come: the exact converse of the current situation of excess world liquidity and low interest rates? If that happens, the task of urban policy makers and central bankers alike will become that much more difficult. The efficient intermediation of financial savings within countries, and across countries, will therefore be as important for urban development as for financial market development per se and for monetary policy makers in the years to come.

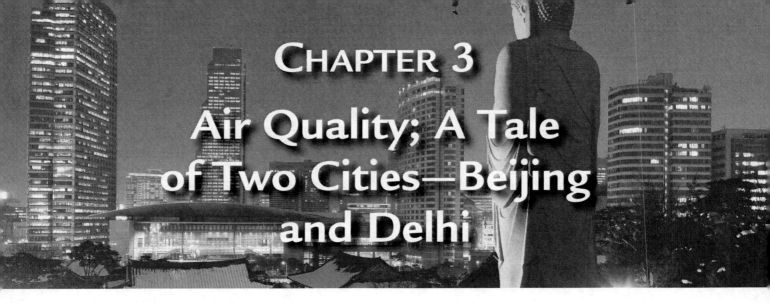

CHAPTER 3
Air Quality; A Tale of Two Cities—Beijing and Delhi

A Toxic Issue

Jessica Sequeira

The 2008 Summer Olympics in Beijing focused environmental attention on pollution in China. But escaping the spotlight was Asia's other tiger, India, where high pollution levels are at least indirectly responsible for tens of thousands of deaths a year. Its capital city is the most egregious offender—according to a 2007 report by the Central Pollution Control Board, Delhi has the highest pollution levels of any major city in the country. Moreover, the Centre for Science and Environment reports that Delhi's air quality has deteriorated sharply over the past two years.

The statistics are alarming. Two of every five residents of New Delhi suffer from a respiratory disease, and hospitalization rates are on the rise for pollution-induced conditions like asthma, lung disease, chronic bronchitis, and heart damage. Researchers have also found evidence that long-term exposure to the microscopic dust particles resulting from pollution can lead to weakened immune systems and lung cancer. Because 70 percent of Delhi's air pollution comes from vehicles, those who work outdoors, including blue-collar workers and taxi drivers, are the most affected group.

The exponential growth of India's economy over the last few decades is responsible for much of the problem. In the 1990s, the loosening of government regulations on foreign trade led to an influx of foreign capital. Many businesses in Silicon Valley and other tech loci began relocating their information technology services and call centers to India. This opportunity for increased prosperity has in essence created a new middle class with Westernized tastes and a budget to match. This hugely ballooning demographic is responsible for 39 percent of Indian consumption now; by 2025, this number will rise to 70 percent.

The direct consequence is the presence of more cars on the roads than ever before. From 1997 to 2006, the number of registered vehicles in New Delhi rose from 1.5 million to 4.5 million, and it continues to increase by an average of 963 private vehicles every day. Tata Motors also plans to begin selling its new "People's Car" in late 2008 at the rock-bottom price of US$3,000. Market analysts predict that the car could expand the Indian car market by as much as 65 percent, with a corresponding increase in emissions. While the increased demand for cars is a tangible indicator of India's continued economic growth, it will challenge India's already gridlocked infrastructure and further increase levels of pollution and congestion.

The government has attempted to tackle the problem of air pollution in the past. A law ordering that all public transportation vehicles use compressed natural gas by 2001 resulted in a noticeable decrease in air pollution. Air quality, by one measure, improved nearly 60 percent from 1994 to 2005. However, recent increases in the number of private vehicles on the road, along with projected growth of the Indian car market, are counteracting this progress. Pollution is steadily creeping back up to pre-2000 levels, and city traffic continues to belch out over 2,000 tons of waste a day.

Yet there is still hope for environmental reforms. If current trends continue, by 2010 the majority of vehicles on the road will be fueled by diesel, a heavy emitter of smoke, particles, and nitrogen oxides. The government could use incentives like taxation to encourage the switch to petrol-fueled cars—which, although far from perfect, emit one-fifth of the pollutants. In addition, India must remedy its severe lack of infrastructure so that traffic is reduced and cars avoid wasting gas while idling. Further development of New Delhi's Mass Transit System, which opened in 2002, could also help cut down on congestion. Currently, however, it only travels to three parts of the city and is in the midst of long-term construction that will not be completed until 2010.

There are yet more hurdles to overcome. Previous attempts at cutting down on major sources of air pollution—for example, a 1990 regulation shifting heavily polluting industrial plants out of New Delhi and a 1997 promise to phase out vehicles over 15 years old—failed in the midst of election season and protests from drivers. Even when such measures are passed, enforcement at the local level is problematic. Notorious corruption in the New Delhi police system makes it easy for big businesses and stubborn drivers to bribe their way out of trouble.

However, for the government to ignore the problem would be a huge oversight. India is understandably committed to technology and industrialization, as it forges its new place on the global economic stage. In the last few years, India has already begun to make much-needed reforms in helping to curb pollution, including the introduction of countrywide emissions norms to toughen up vehicle certification requirements. Conferences and public releases on the topic of air pollution abound. The future thus offers an opportunity to carry forward the programs being put in place today, as well as to act on the promises and possibilities offered by civil-minded scientists and politicians for tomorrow. When it comes to air pollution, India faces the challenging but vital task of cleaning up its act.

Air Pollution

China's Public Health Danger

The rapid pace of economic growth in the Asia Pacific region has been accompanied by resource deple-tion and environmental degradation. Air is the first element to get tainted by industrialization, with air quality becoming an increasingly important public health issue. China in particular has grappled with air pollution and frequent bouts of "haze" in recent years. Chinese smog has been recorded not only in Hong Kong but as far afield as America and Europe. Spire takes a look at how this impacts govern-ment, business and society.

AIR RAID

The Great Smog of London in December 1952 caused over 4,000 deaths, a result of rapid industri-alization and urbanization. Dr Robert Waller, who worked at St Bartholomew's Hospital in the early 1950s, believed that a shortage of coffins and high sales of flowers were the first indications that many people were being killed. "The number of deaths per day during and just after that smog was three to four times the normal level," he said.

The smog, which lasted for five days, was so bad that it infiltrated hospital wards. The death toll was not disputed by the authorities, but exactly how many people perished as a direct result of the fog is unknown. Many who died already suffered from chronic respiratory or cardiovascular complaints. Without the fog, though, they might not have died so early. Mortality from bronchitis and pneumonia increased more than seven-fold as a result of the fog (see Figure 3.1).

In an attempt to remedy the problem, tall smokestacks were built. However, the smokestacks only served to divert the contaminated air to the lakes and forests of Scandinavia in the guise of acid rain.

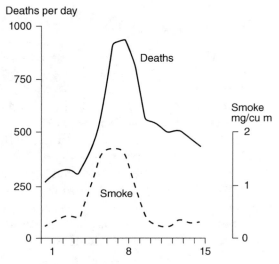

Figure 3.1 Death Rate with Concentrations of Smoke.
Source: Met Office, UK

It became apparent that air pollution could not be addressed by simply deflecting the route of the smoke. As long as the root causes persisted, the problem would be manifested, albeit in varying forms. This realization finally prompted the government to introduce legislation to cut smoke emissions.

Half a century has passed since the Great Smog of London, yet the problem of global air pollution is actually worsening, brought about by industrialization and the growth of the emerging economies which house more than half of the world's population and account for 28 percent of global economic activity[1].

Clean air is an important prerequisite for sustainable economic development and is a basic requirement for human health and welfare. In addition, pollutants contribute to atmospheric problems such as acidification and global climate change, which have impacts on crop productivity, forest growth, biodiversity, buildings and cultural monuments.

WHO is today challenging governments around the world to improve air quality in their cities in order to protect people's health. The health consequences of exposure to polluted air are considerable—approximately 20 to 30 percent of all respiratory diseases appear to be caused by air pollution[2].

AIR TODAY, GONE TOMORROW

Some studies suggest that exposure to air pollution, from living near busy roads for example, is linked to respiratory problems that include reduced lung function, greater prevalence of asthma and respiratory symptoms such as wheezing, and even cardiovascular diseases.

Researchers have found that people living in cities face an increased risk of dying from a heart attack as a result of long-term exposure to air pollution. This increased risk was found to be as large as that associated with being a former smoker. These findings have been reported in the journal Circulation, published by the American Heart Association[3].

Another 1994 report on the adverse effects of particulate air pollution, published in the *Annual Reviews of Public Health,* noted that for each 10 mg/m^3 increase in particulate matter, cardiovascular mortality increased 1.4 percent and respiratory mortality increased at twice that level.

Premature deaths caused by air pollution are estimated at two million worldwide per year. More than half of this burden is borne by people in developing countries.

URBAN AIR POLLUTION IN ASIA

Air pollution is a chronic problem in many of Asia's megacities (ie. cities with a population of over ten million), such as Beijing, Delhi and Jakarta. Coal combustion in factories and power plants and the use of coal and wood for cooking and home heating are primarily to blame. However, automobiles are also an increasingly important contributor to air pollution in much of the world, with more than 600 million vehicles in use, a number that could double over the next 25 years.

A study by WHO has found that 12 of the 15 cities with the highest levels of particulate matter and half a dozen with the highest levels of sulphur dioxide are in Asia. In many countries in the region, the ambient concentration levels of suspended particulate matter and sulphur dioxide exceed WHO standards. Premature mortality and respiratory disease caused by poor air quality have been documented in 16 large metropolitan centres in the region (see Table 3.1).

The problem is increasingly manifesting itself across national boundaries. Spiking air pollution in Asia has changed the atmosphere over the North Pacific. It is causing stronger-than-usual thunderstorms in winter and may even have wider effects on the global climate, according to a study published

---

[1] United Nations Environment Program, Global Environment Outlook 2000.
[2] Air Pollution in the Megacities of Asia (APMA) official website, 2007.
[3] Medical News Today, Air pollution's impact on the heart is as bad as having been a smoker, 16 Dec 2003.

Table 3.1   Impact of Air Pollution in Large Asia Pacific Cities

Health Benefits of Reducing Air Pollution in Large Asia Pacific Cities

Country	City	Population (millions)	Premature deaths (thousands)	Chronic bronchitis cases (thousand)	Respiratory symptoms (millions)	Health benefits from better air quality as a share of urban income (per cent)
China	Beijing	7.0	10.3	81	270	28
	Chengdu	3.0	3.5	29	92	22
	Chongqing	4.0	6.3	44	172	30
	Guangzhou	3.8	2.0	16	51	10
	Harbin	3.1	4.0	34	102	24
	Jinan	2.5	5.0	38	135	38
	Shanghai	9.0	3.8	28	105	8
	Shenyang	4.0	4.9	38	129	23
	Tianjin	5.0	5.7	43	151	21
	Wuhan	4.0	2.0	17	51	9
	Xi'an	3.0	4.1	35	106	26
Indonesia	Jakarta	9.7	6.3	47	142	12
Korea, R. of	Seoul	11.3	2.4	24	72	4
Malaysia	Kuala Lumpur	1.5	0.3	4	11	4
Philippines	Manila	9.7	3.8	33	98	7
Thailand	Bangkok	7.5	2.8	28	82	7
Subtotal	China	48.4	51.6	403	1364	20
	Other countries	39.7	15.6	136	405	7
Total		88.1	67.2	539	1769	

Source: Hughes 1997.

in the "Proceedings of the National Academy of Sciences"[4]. Pollution from China has been picked up in Lake Tahoe in the mountains of eastern California[5]. Satellite data has also tracked China's air pollution drifting towards Korea and Europe.

Being the world's most populous country and its fastest growing manufacturer, China's environmental problems are being closely watched by the international community.

WHERE THERE'S SMOKE, THERE'S FIRE—CHINA'S BATTLE WITH AIR POLLUTION

"If I work in your Beijing, I would shorten my life by at least five years," Zhu Rongji told city officials when he was Prime Minister in 1999. A year before, Beijing was announced the second runner-up for most polluted city in the world[6]. In 2005, satellite data revealed Beijing as air pollution capital of world[7]. China had 16 of the world's 20 most polluted cities in 2006, according to the World Bank.

---

[4] News Target Online, Asian pollution levels impact severity of Pacific storms, March 2007.
[5] The New York Times, 2007.
[6] Asian Economic News, Dec 2001.
[7] The Guradian, UK, Satellite data reveals Beijing as air pollution capital of world, 31 October, 2005.

Roughly two-thirds of the greenhouse gases (GHGs) in the earth's atmosphere are a result of coal and petroleum burning. China now burns more coal (one of the most pollutive of fossil fuels) than the US, Europe and Japan combined, as the country's economy expands. And unfortunately, around 70 percent of its mushrooming energy needs are supplied by coal-fired power stations. The pollution they generate, containing sulfur compounds, carbon and other coal byproducts, tend to cause respiratory damage, heart disease and cancer. The price paid for industrialization is the deaths of more than a hundred thousand people from heart and respiratory system diseases each year.

In 2006, China released 25.9 million tonnes of carbon dioxide. Consequently, over half of the 696 cities under the monitoring of the government have suffered from acid rain. It was estimated that acid rain covered one third of China's land area. From 2000 to 2005, the release of sulfur dioxide from steel companies and coal-fired power stations increased by 25 million tonnes, two-thirds higher than the goal set by the government.

The World Bank has concluded that pollution is costing China an estimated ten percent of its annual GDP in direct damage, such as the impact on crops of acid rain, medical bills, lost work from illness, money spent on disaster relief following floods and the implied costs of resource depletion. If nothing is done about the environment, economic growth could grind to a halt.

Certainly, awareness of China's environmental problems is rising among policymakers—reflected in a new package of right-sounding initiatives like a "green GDP" indicator to account for environmental costs. So is the pressure, from both internal and international sources, to fix them. But while all developing economies face this issue, there are historical, political and institutional reasons why resolving the crisis will be a long and complicated process in China.

PUTTING OUT THE FLAMES

Faced with severe air pollution driven by industrialization, the Chinese government is taking a proactive stance to tackle this problem. Its first step was signing the Kyoto Protocol to the United Nations Framework Convention on Climate Change in 2005.

Clean-up measures are now being announced thick and fast. In Beijing, after years of apparent inaction, a total ban on leaded petrol for cars was implemented within the space of just six months.

The authorities are also taking action against polluting factories. Some have been closed, and others are under threat if they do not drastically cut pollution by the end of 2007.

Beijing has plans to phase out coal use as well. "We're changing the energy structure of the whole city", said environment official Li Tiejun[8]. "First we tackled the small food stoves; now, in just two years, almost the whole catering industry has gone over to natural gas or electricity, and all small and medium size industrial boilers are using clean fuels . . . It's like London in the 1950s. Once you stop coal use the problem is solved."

The enthusiasm shown by the authorities has sparked a chain reaction among private companies. For example, the biggest taxi company in Shanghai, Dazhong, uses natural liquefied gas, rather than diesel, to eliminate black emission.

Recently, a 62-page climate change plan was unveiled and officials promised to put the issue at the heart of its energy policies.

There is some cause for optimism but progress on pollution is unlikely to be as rapid or uniform as the government and environmentalists desire.

Good intentions have so far had limited impact on the ground, due to China's vast, decentralized bureaucracy. As Ken Lieberthal, a China expert at the University of Michigan, explains: "Much of the environmental energy generated at the national level dissipates as it diffuses through the multi-layered state structure, producing outcomes that have little concrete effect."

The State Environmental Protection Administration (SEPA), the government's enforcement agency in the fight against pollution, is under-resourced with little funds and just a few hundred central staff.

---

[8] BBC News, China's Environmental Challenge, 17 November 2000.

Around the country, SEPA's branches are tasked to monitor pollution, enforce standards and collect fines. But their salaries and pensions come from local governments—whose priorities are to maintain growth and employment in their jurisdiction—rather than from Beijing. This creates loyalty dilemmas. Typically, a bureau would impose a fine on a pollutive local enterprise but then pass the money on to the local administration, which refunds it to the company via a tax break. SEPA's impotence is the main reason why penalties, even when it can impose them, are absurdly light.

Chinese leaders have set tough new targets to reduce the use of energy per unit of economic output by 20 percent and pollution by 10 percent, between 2006 and 2010. However, China now accounts for almost half of the world's flat glass and cement production, more than a third of steel output and nearly as much of aluminium. Heavy industry consumes 54 percent of China's energy, up from 39 percent five years ago[9]. The rise of heavy industry explains why China failed to meet its pollution-cutting targets in 2006 and will find it hard to do so by the end of the decade.

In the final analysis, the Chinese government is reluctant to slow its rate of economic growth too much in order to meet environmental goals. In this regard, its stance is no different from that of most of the world's governments today. Moreover, it rightly points out that much of its air pollution is generated by the manufacture of cheap exports to developed countries, who should therefore bear part of the responsibility and costs.

Ma Kai, the chairman of the National Development and Reform Commission—the chief economic planning agency which also handles climate change—commented on China's stand when releasing the first national plan to combat global warming. He considered a country's stage of development, contribution to cumulative greenhouse gases in the atmosphere and per capita emissions the proper benchmarks. Measured by such indicators, China's contribution to the present problem was relatively small.

"Possible future emissions should not be used as an excuse to ask developing countries to undertake cuts and to do so in a way that is too early, too abrupt and too blunt," he said. "The ramifications of limiting the development of developing countries would be even more serious than those from climate change."[10]

A BIG GREEN OPPORTUNITY

But there is hope yet, as demonstrated by foreign companies flocking to China to build and sell environmental technologies. Vendors making everything from water purifiers to wind turbines and hydrogen-powered vehicles are flocking to China to sell their wares and services.

Business-to-Business

Niche player Fuel Tech, global engineering firms Foster Wheeler and McDermott international, and industrial behemoths like General Electric (GE) are all positioning themselves to sell more low-emissions electric generation parts to China[11].

"If you're in the air pollution control business, you go to where the need is," said John Norris, chief executive of Fuel Tech, a company that makes products to reduce plants' nitrogen oxide emissions.

The National Development and Reform Commission (NDRC) has disclosed China's 15-year goal of shifting to renewable energy: by 2020, renewable energy sources are targeted to reach 16 percent of total energy capacity[12]. To reach this ambitious goal, the NDRC signed agreements in July with

---

[9] Financial Times online, China pollution fuelled by heavy industry, 1 May 2007.
[10] Financial Times Limited 2007: China urges rich nations to lead on climate, 4 June 2007.
[11] DowJones MarketWatch, China's smog + power demand = sales opportunity, June 2007.
[12] WordWatch Institute, China Needs to Move Quickly on Energy Savings, November 2006.

all provincial and municipal governments, as well as with 14 major state companies, laying out their specific responsibilities for energy saving. Industries involving hydro-energy, wind energy, solar energy, biomass energy and geothermal energy are highly-prepared for development.

China is currently encouraging leading companies like GE Power and Veolia of France to market technologies that will harness the methane gas produced from decomposing garbage and sewage, as well as the huge amounts of gas that escape from China's coal mines.

The other half of the air pollution equation is reducing harmful emissions from traditional energy usage. On this front, the China government has set ambitious targets.

Desulphurization technologies and facilities currently have a strong market in China, thanks to government initiatives. In 2005, the authorities unveiled a detailed mandate mapping out how thousands of China's coal-fired power plants would be "desulphurized." The mandate encouraged power plants to equip themselves with desulphurization facilities and promised to defray the high costs.

One of the McDermott group's business units, Babcock & Wilcox, started supplying the Chinese market with industrial boilers in 1945 and large coal-fired steam generators in the 1980s. More recently, it entered the pollution abatement business by licensing its wet-fuel gas-desulphurization technology.

Foster Wheeler, a company that derives it revenues mainly from huge infrastructure projects, is selling Chinese power companies a type of boiler that cuts emissions of nitrogen oxide.

Indoor pollution could be an even greater contributor to respiratory diseases than outdoor pollutants. An estimated 64 percent of China's population use coal in their homes, and about 22 percent of rural homes rely on coal for domestic fuel[13]. China has, in fact, initiated ongoing programs to address the problems caused by household coal use. The government has also taken steps to promote energy-efficiency in buildings.

But it is likely to take years before China's appetite for cleaner power translates into sizeable earnings for firms that supply emissions-reducing power parts, say analysts. Among the hurdles, managers of China's centrally-planned economy are constantly weighing the increased costs of pollution controls against electricity demand. And sales to utilities take time.

Business-to-Consumer

Short-term profits from fighting air pollution will mainly accrue in the consumer realm. For the second year in a row, an annual survey in 2006 showed that pollution was still the top complaint of expatriates in China[14].

In China, there is a strong emphasis on preventative medicine—there is a large market for non-prescription products that promote overall health. The health supplement and pharmaceutical markets will benefit from preventive products and symptomatic treatments for respiratory diseases.

The pharmaceutical market in China is currently worth an estimated US$33.5 billion[15], a profitable market that global pharmaceutical powerhouses can attest to (see Table 3.2).

The skincare industry, too, benefits from anti-oxidizing products. Cosmetics giant L'Oreal launched an initiative to fight pollution's effects on skin and hair back in 2003. The company's total sales reached US$180m in China in that year[16]. "The growth in sales in the Chinese market was the highest of all L'Oreal companies in various parts of the world," said Paolo Gasparrini, President of L'Oreal (China).

---

[13] Environmental Health Perspectives, On a Different Scale: Putting China's Environmental Crisis in Perspective, October 2000.
[14] ChinaCSR.com, Annual Survey Shows Pollution Still Top Gripe of Expatriates in China, 2 February 2007.
[15] Alberta Hong Kong Office official website, 2007.
[16] People's Daily Online, L'Oreal sales reached US$180m in China last year, March 2004.

Table 3.2   China's Biggest Drug Sellers

Company	Sales (USD)*	YOY Growth*
Yangzijiang Pharmaceutical	$215 million	38%
Pfizer	$160 million	26%
AstraZeneca	$133 million	32%
Roche	$130 million	32%
Novartis	$126 million	14%
Fresenius	$112 million	18%
GlaxoSmithKline	$103 million	—
*rolling 12 month period ending July 31, 200		

Source: IMS Health.

I DO MY PART, YOU DO YOURS

The China government clearly "gets it" when it comes to the environment. Numerous laws, regulations and targets have been issued by the government to address environmental issues and air pollution in particular. However China's current weak enforcement infrastructure means that its laws make too little impact on the ground. But there are signs that this will change in the coming years.

International agencies are increasingly tying funds to environmental criteria, while foreign governments are beginning to complain about China's dust storms and GHG emissions. An even bigger factor in accelerating change is the 2008 Olympics. Described as China's "coming-out party to the world", the Olympics have led to Beijing's government moving out factories and introducing clean-vehicle technology. In fact, the city has introduced natural gas in 2,100 buses and plans to have as many as 8,000 of such environmentally-friendly vehicles on the roads by the time the Olympic torch gets lit[17].

On a macro scale, the Chinese Academy of Engineering and State Environmental Protection Administration has taken the lead to develop a study to explore and identify strategic guidelines, priorities and measures on the environmental protection of China. Formally launched in Beijing on May 11 2007, the implementation of this project is cited as a key step towards building an environmentally-friendly society.

Although China has promised to put climate change at the heart of its energy policies, it says developed countries have an "unshirkable responsibility" to take the lead on the issue by cutting emissions. At the end of the day, China will make rapid progress in combating air pollution; it will improve the enforcement of laws and targets that have been promulgated by the authorities and this will provide growing opportunities to clean air technology vendors; however the pace of improvement will be dictated by the extent to which the world's rich nations are prepared to help shoulder the economic costs.

---

[17] Business Week, Cleaning up in China, June 2005.

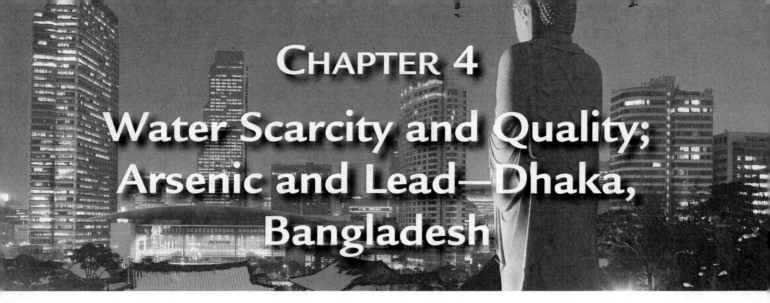

CHAPTER 4
Water Scarcity and Quality; Arsenic and Lead—Dhaka, Bangladesh

Groundwater Arsenic Contamination in Bangladesh: Causes, Effects and Remediation

Md. Safiuddin Department of Civil and Environmental Engineering
Md. Masud Karim Engconsult Limited

ABSTRACT

The serious arsenic contamination of groundwater in Bangladesh has come out recently as the biggest natural calamity in the world. The people in 59 out of 64 districts comprising 126,134 km² of Bangladesh are suffering due to the arsenic contamination in drinking water. Seventy five million people are at risk and 24 million are potentially exposed to arsenic contamination. Most of the recognized stages of arsenic poisoning have been identified in Bangladesh and the risk of arsenic poisoning in the population is increasing everyday. The severity of arsenic contamination is demanding extensive research in this field. Many studies have been carried out in Bangladesh, in West Bengal, India and other countries as well, but the situation is still out of sound control. The present study is an overview of groundwater arsenic contamination in Bangladesh. This study highlights the causes and mechanisms of arsenic contamination in groundwater. The effects of arsenic contamination on human health have been revealed. It also presents several measures to remedy the arsenic contamination in groundwater.

INTRODUCTION

Groundwater arsenic contamination in Bangladesh is reported to be the biggest arsenic calamity in the world in terms of the affected population[1]. The Government of Bangladesh has addressed it as a national disaster. Arsenic contamination of groundwater in Bangladesh was first detected in 1993[2].

Table 4.1   Percentage of Groundwater Surveyed in 1998 by British Geological Survey with Arsenic Levels above 0.05 mg/l[4]

District	Percentage of Groundwater Surveyed	District	Percentage of Groundwater Surveyed
Bagerhat	66	Madaripur	93
Barisal	63	Magura	19
Brahmanbaria	38	Manikganj	15
Chandpur	96	Meherpur	60
Chittagong	20	Moulvibazar	12
Chuadanga	44	Munshiganj	83
Comilla	65	Narail	43
Cox's Bazar	3	Narayanganj	24
Dhaka	37	Nawabganj	4
Faridpur	66	Noakhali	75
Feni	39	Pabna	17
Gopalganj	94	Pirojpur	24
Jessore	51	Rajbari	24
Jhalakati	14	Rajshahi	6
Jhenaidah	26	Satkhira	73
Khulna	32	Shariatpur	80
Kushtia	28	Syllhet	19
Lakshmipur	68		

Further investigations were carried out in the following years. The institutions that contributed in the investigations are the School of Environmental Studies (SOES) from Jadavpur University in Calcutta, Bangladesh Atomic Energy Commission (BAEC), Dhaka Community Hospital (DCH), Department of Public Health Engineering (DPHE), and National Institute of Preventive and Social Medicine (NIPSOM). DPHE collected and analyzed 31,651 well water samples with the assistance of WHO, UNICEF and DFID[3]. The laboratory reports have confirmed that the groundwater in Bangladesh is severely contaminated by arsenic. The millions of shallow and deep wells that had been sunk in various parts of the country are dispensing their own special brand of poison. In consequence, a large number of populations in Bangladesh are suffering from the toxic effects of arsenic contaminated water.

Recent studies in Bangladesh indicate that the groundwater is severely contaminated with arsenic above the maximum permissible limit of drinking water. In 1996, altogether 400 measurements were conducted in Bangladesh[4]. Arsenic concentrations in about half of the measurements were above the maximum permissible level of 0.05 mg/l in Bangladesh. In 1998, British Geological Survey (BGS) collected 2022 water samples from 41 arsenic-affected districts[4]. Laboratory tests revealed that 35% of these water samples were found to have arsenic concentrations above 0.05 mg/l. The survey results are shown in Table 4.1.

NIPSOM and SOES conducted a study in Rajarampur village of Nawabganj district in 1996. The report shows that 29% of the 294 tube-wells tested had arsenic concentrations greater than 0.05 mg/l[5]. Between September 1996 and June 1997, DCH also conducted a field survey in Samta village of Jessore district in collaboration with SOES[6]. In total, 265 tube-wells were examined and it was found that about 91% of the tube-wells had arsenic concentrations higher than 0.05 mg/l. Further studies by SOES and DCH in the Ganges delta exhibited that 59% of the 7800 groundwater samples

Table 4.2 Statistics of Arsenic Calamity in Bangladesh[10]

Total Number of Districts in Bangladesh	64
Total Area of Bangladesh	148,393 km^2
Total Population of Bangladesh	125 million
GDP Per Capita (1998)	US$260.00
WHO Arsenic Drinking Water Standard	0.01 mg/l
Bangladesh Arsenic Drinking Water Standard	0.05 mg/l
Number of Districts Surveyed for Arsenic Contamination	64
Number of Districts Having Arsenic above 0.05 mg/l in Groundwater	59
Area of Affected 59 Districts	126,134 km^2
Population at Risk	75 million
Potentially Exposed Population	24 million
Number of Patients Suffering from Arsenicosis	8,500
Total Number of Tube-wells in Bangladesh	4 million
Total Number of Affected Tube-wells	1.12 million

had arsenic concentrations greater than 0.05 mg/l[7]. So far from August 1995 to February 2000, SOES and DCH had jointly analyzed 22003 tube-well water samples collected from 64 districts in Bangladesh[8]. Five years sampling results indicate that out of 64 districts in Bangladesh, arsenic in groundwater is above 0.01 mg/l in 54 districts and above 0.05 mg/l in 47 districts. The experts from Bangladesh Council for Scientific and Industrial Research (BCSIR) have been found the highest level of arsenic contamination, 14 mg/l of shallow tube-well water in Pabna[9]. The recent statistics on arsenic contamination indicate that 59 out of 64 districts of Bangladesh have been affected by arsenic contamination. Approximately, arsenic has contaminated the ground water in 85% of the total area of Bangladesh and about 75 million people are at risk[10]. It has been estimated that at least 1.2 million people are exposed to arsenic poisoning. The reported number of patients seriously affected by arsenic in drinking water has now risen to 8500[11]. As the people are getting arsenic also from food chain such as rice, fish and vegetables, the problem is growing more severe. The current statistics of arsenic calamity given in Table 4.2.

In fact, the severity of arsenic contamination has caused a serious panic for the people in Bangladesh. It is felt that the magnitude of arsenic problem in Bangladesh surpasses the aggregate problem of all the twenty countries of the world where groundwater arsenic contamination has been reported. This is the worst case of mass poisoning the world has ever experienced. Alarm bells are now ringing in Bangladesh since arsenic in groundwater has emerged as a serious problem across the country. The problem is made more complex by the fact that the contamination is occurring below the ground where it cannot be easily identified.

CAUSES OF ARSENIC CONTAMINATION

Intermittent incidents of arsenic contamination in groundwater can arise both naturally and industrially. The natural occurrence of arsenic in groundwater is directly related to the arsenic complexes present in soils. Arsenic can liberate from these complexes under some circumstances. Since arsenic in soils is highly mobile, once it is liberated, it results in possible groundwater contamination.

The alluvial and deltaic sediments containing pyrite has favored the arsenic contamination of groundwater in Bangladesh. Most regions of Bangladesh are composed of a vast thickness of alluvial and deltaic sediments, which can be divided into two major parts—the recent floodplain and the terrace areas. The floodplain and the sediments beneath them are only a few thousand years old. The terrace areas are better known as Madhupur and Barind Tracts and the sediments underlying them are much older than the adjacent floodplain. Most of the arsenic is occurring in the younger sediments derived from the Ganges Basin. The investigators found that there is a layer containing arsenic compound at a depth of 20 to 80 meters[12]. This layer is rich in arseno-pyrite, pyrite, iron sulfate, and iron oxide as revealed by the geological investigation. The researchers also inferred that, although arsenic is occurring in the alluvial sediments, the ultimate origin of arsenic is perhaps in the outcrops of hard rocks higher up the Ganges catchment. These outcrops were weather-beaten in the recent geological past and then the eroded soil was deposited in West Bengal and Bangladesh by the ancient courses of the Ganges[13]. Arsenic in sediment or water can move in adsorbed phase with iron, which is available in plenty in the Himalayas. Here about 100 to 300 mg/kg arsenic combined with iron oxides can be found in the sediments under aerobic conditions[14]. When these sediments were deposited in Bengal basin under tidal environment, it came under anaerobic condition. The sulfate available in Bengal basin was reduced to hydrogen sulfide in presence of sulfur reducing bacteria. Iron minerals and hydrogen sulfide rapidly tie together to form iron sulfide. Arsenic had been absorbed on the surface of iron sulfide and produced arsenopyrite. This mineral usually remains stable unless it is exposed to oxygen or nitrate. In aerobic environment, arseno-pyrite is oxidized in presence of oxygen and arsenic adsorbed with iron sulfide becomes mobilized.

The groundwater in Bangladesh has declined progressively due to the excessive extraction of water for irrigation and domestic water supply, lack of water management and inadequate recharge of the aquifer. The groundwater declined beyond 8 meters in 12% areas of Bangladesh in 1986. This extent rose to 20% areas in 1992 and 25% areas in 1994[15]. The study on forecasting groundwater level fluctuation in Bangladesh indicated that 54% areas of Bangladesh are likely to be affected up to 20 meters in some areas particularly in northern part of the country. Excessive groundwater extraction may be the vital reason for creating a zone of aeration in clayey and peaty sediments containing arseno-pyrite. Under aerobic condition, arseno-pyrite decomposes and releases arsenic that mobilizes to the subsurface water. The mobilization of arsenic is further enhanced by the compaction of aquifers caused by groundwater withdrawal.

MECHANISM OF ARSENIC CONTAMINATION

Presently, there are two well-known theories about the mechanism of arsenic contamination in groundwater. These are oxidation and oxyhydroxide reduction theory. The oxidation theory is so far the accepted theory. According to this theory, arsenic is released from the sulfide minerals (arseno-pyrite) in the shallow aquifer due to oxidation[16]. The lowering of water table owing to over exploitation of groundwater for irrigation has initiated the release of arsenic. The large-scale withdrawal of groundwater has caused rapid diffusion of oxygen within the pore spaces of sediments as well as an increase in dissolved oxygen in the upper part of groundwater. The newly introduced oxygen oxidizes the arseno-pyrite and forms hydrated iron arsenate compound known as pitticite in presence of water. This is very soft and water-soluble compound. The light pressures of tube-well water break the pitticite layer into fine particles and make it readily soluble in water. Then it seeps like drops of tea from the teabag and percolates from the subsoil into the water table. Hence, when the tube-well is in operation, it comes out with the extracted water. This mechanism is portrayed in Figure 4.1.

The alternative hypothesis on the arsenic contamination is the oxyhydroxide reduction theory proposed by Nickson et. al.[17]. This theory has been accepted recently by some scientists and researchers as main the process for mobilization of arsenic in groundwater. According to this theory, arsenic is derived by desorption from ferric hydroxide minerals under reducing conditions. Ferric hydroxide minerals are present as coatings in the aquifer sediments. In anaerobic groundwater, these sedimentary minerals release its scavenged arsenic.

The oxidation of arseno-pyrite could be the main mechanism for the groundwater arsenic contamination in Bangladesh but there is not enough hydrological and geochemical data to validate the process completely. The validity of oxyhydroxide reduction theory is also questionable

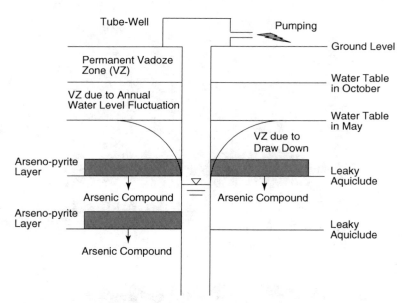

Figure 4.1   Aerobic Condition in Groundwater around a Heavy-Duty Tube-Well[13].

due to the lack of comprehensive sampling and systematic analysis of iron oxy-hydroxides in the affected areas.

EFFECTS OF ARSENIC CONTAMINATION

Effects on Human Health

The data collected by the governmental bodies, NGOs and private organizations reveal that a large number of populations in Bangladesh are suffering from melanosis, leuco-melanosis, keratosis, hyperkeratosis, dorsum, non-petting oedema, gangrene and skin cancer[18]. Melanosis (93.5%) and keratosis (68.3%) are the most common presentations among the affected people. Patients of Leucomelanosis (39.1%) and hyper-keratosis (37.6%) have been found in many cases. Few cases of skin cancer (0.8%) have also been identified among the patients seriously affected by the arsenicals (arsenite and arsenate). Figure 4.2 and Figure 4.3 show skin lesions on palm and soles, respectively.

The occurrence of arsenic diseases depends on the ingestion of arsenic compounds and their excretion from the body. It has been reported that 40% to 60% arsenic can be retained by the human body[19]. It indicates that the level of hazards will be higher with the greater consumption of arsenic contaminated water. The daily consumption of arsenic contaminated water is very high in Bangladesh, especially in villages. The villagers consume about five liters water per day due to manual labor. Moreover, they consume plenty of rice-water and all of their foods are also cooked using arsenic polluted water. Therefore, the people of villages in the affected areas are getting more arsenic than expected. So far SOES and DCH[8] had analyzed 11000 hair, nail, urine and skin-scale samples collected from the affected villages in Bangladesh. The analysis shows that around 90% of people have arsenic in their hair, nail and urine above the normal level. The normal concentration of arsenic in hair is 0.08–0.25 mg/kg and 1 mg/kg indicates the toxic level[20]. The normal arsenic content in nails is 0.43–1.08 mg/kg[21] and the normal amount of arsenic in urine ranges from 0.005 to 0.040 mg/day[20]. Table 4.3 shows that the arsenic contents in hairs, nails, urine and skin scales of the affected people are very high in Bangladesh.

There are several factors may have been responsible for triggering off the arsenic-related diseases in Bangladesh. The primary reason appears to be the malnutrition, a state that describes 80 percent of the population of Bangladesh. Having less immunity, a huge number of people are suffering from

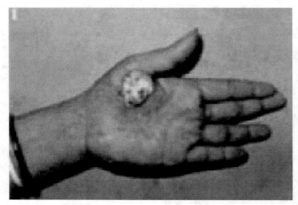

Figure 4.2   Skin Lesions on Palm due to Arsenic Intake in Drinking Water[16].

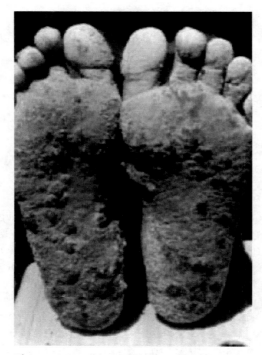

Figure 4.3   Skin Lesions in Soles due to Arsenic Intake in Drinking Water[4].

the chronic arsenic poisoning. Many People have died, many are dying and many will die of arsenic diseases. In brief, the majority of the people in Bangladesh are grappling with the massive health crisis caused by the arsenic diseases.

Social Effects

Although what is causing arsenic contamination in groundwater is not clear indisputably, its effect on people is well known. The sudden increase in arsenic related diseases has panicked the local people. The native people consider the arsenic diseases contagious. In many instances, the people suffering from arsenic diseases have been ostracized by neighbors, friends and relatives. The affected people are either avoided or discouraged to appear in public places. The affected children are often barred from attending schools and the adults are discouraged from attending offices and any public

Table 4.3   Survey Results of Arsenic Contents in Hair, Nails, Skin Scales and Urine[8]

Field Survey from August 1995 to February 2000 (239 Days)	
Total Hair Samples Collected from 210 Arsenic Affected Villages	4386
Percentage of Samples Having Arsenic above Toxic Level	83.15%
Total Nail Samples Collected from 210 Arsenic Affected Villages	4321
Percentage of Samples Containing Arsenic above Normal Level	93.77%
Total Urine Samples Collected from 20 Arsenic Affected Villages	1084
Percentage of Samples Having Arsenic above Normal Level	95.11%
Total Samples of Skin Scales	705
Percentage of Samples Containing Arsenic above Toxic Level	97.44%
Field Survey from April 1999 to February 2000 (27 Days)	
Total Hair Samples	1054
Percentage of Hair Samples Having Arsenic above Toxic Level	89.35%
Total Nail Samples	1000
Percentage of Nail Samples Containing Arsenic above Normal Level	94%
Total Urine Samples	41
Percentage of Urine Samples Having Arsenic above Normal Level	97.50%
Total Samples of Skin Scales	115
Percentage of Samples Containing Arsenic above Toxic Level	100%

meetings. Qualified persons are refused jobs when found suffering from arsenicosis. Those affected with a higher level of contamination are considered incapable of working and hence victimized by the growing poverty. The situation is worse for women. The women suffering from arsenic diseases are increasingly facing ostracization and discrimination. Young women suffering from arsenicosis are often compelled to stay unmarried. Married women affected by arsenic are no longer considered acceptable as wives due to skin lesions and sent back to their parents with children. Thus, the unaffected parents and children are also suffering socially with the affected females. Above all, the affected people are losing their as usual social relation with the neighbors and relatives.

REMEDIATION

The situation of groundwater arsenic contamination in Bangladesh is so serious that the immediate steps should be taken to find and deliver adequate potable water to all seriously affected areas for drinking and cooking purposes. The following measures should be implemented depending on the cost effectiveness:

1. Alternative sources of drinking water: innovative alternative sources such as pond sand filters, infiltration galleries, or Ranney wells, and in some places even rainwater harvesting can be adopted to alleviate the arsenic disaster.
2. Use of surface water: Existing surface water could be purified by filtration and chlorination, and even by ultraviolet disinfection or solar radiation and can be used in drinking and other house hold purposes.
3. Removal of arsenic by chemical precipitation: Coagulants such as the salts of aluminium and iron should be used to remove the arsenic from domestic drinking water.
4. Removal of arsenic by oxidation: Oxidants such as free chlorine, ozone, permanganate, hypochlorite, and Fenton reagent (H_2O_2/Fe^{2+}) should be used to remove arsenic from drinking water.

5. Extraction and distribution of arsenic free groundwater from deep aquifers: If other alternatives are costly and complicated potable drinking water can be extracted and distributed from deep aquifers.
6. Removal of arsenic from water collected from the existing contaminated sources by filtration: Water filters should be used at drinking water treatment plant or at each individual household source.
7. Removal of arsenic from the existing water sources: The sources of arsenic contamination must be controlled and arsenic contaminated soil and shallow groundwater aquifers should be cleaned to prohibit the future contamination.
8. In-situ remediation of arsenic contaminated groundwater: This can be achieved by using iron filings permeable walls.
9. Implementation of efficient water supply system: A safe and long lasting efficient water supply system should be implemented for the whole country.
10. Development of sewage and waste disposal system: An efficient sewage and waste disposal system should be developed to prevent the contamination of soil and water supplies.

Principally, the best solution appears to be the restoration of natural river flow and groundwater level. The natural groundwater level that existed prior to 1975 should be restored. The flushing of arsenic contaminants may take a long time but these will be diluted by the restoration of natural rivers and groundwater aquifers. Thus, the severity of arsenic contamination will be reduced gradually. Besides, this will provide plenty of water for drinking, irrigation, and industry.

CONCLUDING REMARKS

Arsenic contamination is not peculiar to Bangladesh alone. This is a global problem. There are other countries in the world that had experienced or going through this problem. The great difference is the degree and velocity of this environmental disaster in Bangladesh for the number of people at risk is higher than other countries. Even this problem is not as severe as in the neighboring West Bengal, where the similar disaster is taking place. In fact, arsenic contamination is not as severe or as wide spread in anywhere as it is in Bangladesh. Thousands of arsenic affected patients have already been identified. If the people continue to use arsenic contaminated water, millions will lose their health or die within a few decades. Those who will survive are in a danger of carrying genetic diseases to future generation. Unfortunately, the basic facts in Bangladesh are that the people in the affected regions are still unaware of arsenic contamination and its hazardous effects. The governmental efforts are much less than needed to mitigate the crisis. Hence, the immediate involvement of international community is urgent to combat the slow onset disaster and save the poor people.

Economically and technologically, Bangladesh is not in a firm position to solve the arsenic crisis herself. She needs the help of the international community. Environmental experts and funds are desperately needed to save the lives of millions of people affected by deadly arsenic. The international community has the economic resources, environmental experts, and technologies to mitigate the arsenic contamination in groundwater. The support of United Nations, donor countries, donor organizations, agencies, and individuals is essential to save the suffering people from the devastating arsenic disaster.

RECOMMENDATIONS

Although groundwater arsenic contamination in Bangladesh has been declared a national disaster by the government, its seriousness is yet to be fully comprehended. If the following recommendations for research and development are successfully carried out, the remediation of arsenic contamination will be much easier.

1. It is highly desirable to form a research group with geologists, hydrologists, geo-chemists, water supply and environmental engineers, and public health experts to conduct in-depth investigation on the sources and causes of arsenic contamination in groundwater.
2. A comprehensive research plan should be developed to determine the geological, hydrogeological and geochemical factors controlling the chemical reactions generating and releasing arsenic to groundwater.
3. A national groundwater resources management policy be established in order to limit the indiscriminate abstraction of groundwater.
4. It is highly recommended that every donor projects in arsenic mitigation bylaw ensure community participation for smooth running in future.
5. A comprehensive water distribution system should be implemented and an efficient monitoring system should be established to provide potable water and to prevent future arsenic contamination in drinking water.
6. An effective sewage disposal system should also be established to accompany any deployment of water distribution system.
7. Guidelines on the disposal of arsenical wastes should be established to minimize the contamination in soil and water.
8. An estimate of annual arsenic use in agriculture is required and the short-term or long-term environmental impact of arsenic use in cultivation should be assessed.
9. The population exposed to the arsenic contamination should be advised about the arsenic in drinking water, the sources of arsenic-free water, and the importance of compliance with treatment programs including the nutrition.

REFERENCES

1. Talukder, S.A., Chatterjee, A., Zheng, J., Kosmus, W., "Studies of Drinking Water Quality and Arsenic Calamity in Groundwater of Bangladesh", *Proceedings of the International Conference on Arsenic Pollution of Groundwater in Bangladesh: Causes, Effects and Remedies*, Dhaka, Bangladesh, February 1998.
2. Khan, A.W. et.al., "Arsenic Contamination in Groundwater and Its Effect on Human Health with Particular Reference to Bangladesh", *Journal of Preventive and Social Medicine*, Vol. 16, No. 1, pp.65–73, 1997.
3. Daily Star Report, "An Urgent Call to Save a Nation", *The Daily Star*, A national daily newspaper of Bangladesh, 10 March 1999.
4. Smith, A.H., Lingas, E.O., and Rahman M., "Contamination of Drinking- Water by Arsenic in Bangladesh: a Public Health Emergency", *Bulletin of World Health Organization*, Vol. 78, No. 8, WHO, pp.1093–1103, 2000.
5. Ahmad, S.A. et.al., "Arsenic Contamination in Ground Water and Arsenicosis in Bangladesh", *International Journal of Environmental Health Research*, Vol. 7, pp. 271–276, 1997.
6. Biswas, B.K. et.al. "Detailed Study Report of Samta, One of the Arsenic- Affected Villages of Jessore District, Bangladesh", *Current Science*, Vol. 74, pp.134–145, 1998.
7. Chowdhury, T.R., et. al., "Arsenic Poisoning in the Ganges Delta", *Nature*, Vol. 401, pp.545–546, 1999.
8. SOES & DCH, "Summary of 239 Days Field Survey from August 1995 to February 2000", *Groundwater Arsenic Contamination in Bangladesh*, A Survey Report Conducted by the School of Environmental Studies, Jadavpur University, Calcutta, India and Dhaka Community Hospital, Dhaka, Bangladesh, 2000.
9. New Nation Report, "Immediate Government Steps Needed, Millions Affected by Arsenic Contamination", The New Nation, A daily newspaper of Bangladesh, 11 November 1996.
10. www.dainichi-consul.co.jp/english/arsen.htm, "Arsenic Calamity of Bangladesh", *On-line Arsenic Page*, Dainichi Consultant, Inc., Gifu, Japan, 2000.
11. Daily Star Report, "8500 Arsenic Patients Detected in Country", *The Daily Star*, A national daily newspaper of Bangladesh, 11 September 2000.
12. Independent Report, "Remedies for Arsenic Poisoning", *The Independent*, A national daily newspaper of Bangladesh, 16 March 1998.
13. Karim, M.M., Komori, Y., and Alam, M., "Subsurface Arsenic Occurrence and Depth of Contamination in Bangladesh", *Journal of Environmental Chemistry*, Vol. 7, No. 4, pp.783–792, 1997.

14. Mortoza, S., "Arsenic Contamination – Too Formidable a Foe", *On-line Article from West Bengal & Bangladesh Arsenic Information Center*, Water Environment International, Corporate Office in Resource Planning and Management, Dhaka, Bangladesh.

15. NMIDP, "Groundwater Development Potential", *National Minor Irrigation Development Project*, Bangladesh Water Development Board, 1996.

16. Mandal, B.K., Chowdhury, T.R., Samanta, G., Mukherjee, D.P., Chanda, C.R., Saha, K.C., Chakraborti, D., "Impact of Safe Water for Drinking and Cooking on Five Arsenic-Affected Families for 2 Years in West Bengal, India", *The Science of Total Environment*, Vol. 218, pp.185–201, 1998.

17. Nickson, R.T., McArthur, J.M., Ravenscroft, P., Burgess, W.G., and Ahmed, K.M., "Mechanism of Arsenic Release to Groundwater, Bangladesh and West Bengal", *Applied Geochemistry*, Vol. 15, pp.403–413, 2000.

18. Karim, M., "Arsenic in Groundwater and Health Problems in Bangladesh", *Water Resources*, Vol. 34, No. 1, pp.304–310, 2000.

19. Farmer, J.G., and Johnson, L.R., "Assessment of Occupational Exposure to Inorganic Arsenic Based on Urinary Concentrations and Speciation of Arsenic", *Br. J. Ind. Med.*, Vol. 42, pp.342–348, 1990.

20. Arnold, H.L., Odam, R.B., and James, W.D., "Disease of the Skin", *Clinical Dermatology*, 8th Edition, W.B. Saunders, Philadelphia, USA, p.121, 1990.

21. Dhar, R.K., Biswas, B.K., Samanta, G., Mandad, B.K., Chakraborti, D., Roy, S., Jafar, J., Islam, A., Ara, G., Kabir, S., Khan, A.W., Ahmed, S.A., Hadi, A.A., "Groundwater Arsenic Calamity in Bangladesh", *Current Science*, Vol. 73, No. 1, pp.48–59, 1997.

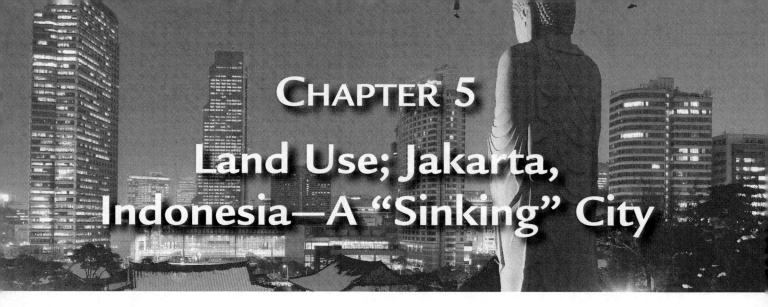

CHAPTER 5
Land Use; Jakarta, Indonesia—A "Sinking" City

Will Jakarta be the Next Atlantis? Excessive Groundwater use Resulting from a Failing Piped Water Network

Nicola Colbran

INTRODUCTION

Historically, piped water in Jakarta was intended to serve selected residents, industries and businesses in accordance with the politics of the era. Whether to segregate between races under the Dutch colonial government, to indicate modernity under first president Soekarno, or to support economic development under second president Soeharto, the supply of piped water has favoured few and left many without access. Those that are connected frequently complain about the low quality of water supplied, the quantity and continuity of the water, the level of water pressure and high water tariffs. Groundwater, on the other hand, has for centuries been a cheap and reliable water source. The result is that many households in Jakarta, as well as industry, business, luxury apartment complexes and hotels do not connect, and are not connected, to the piped water network. They instead use alternative water sources and distribution methods in particular groundwater. The impact of this on groundwater levels and Jakarta's natural environment is unsustainable. Excessive and unlicensed groundwater use is causing significant land subsidence, pollution and salinisation of aquifers, and increased levels of flooding in Jakarta. It is also lowering the water table, making it increasingly difficult for the estimated 70 per cent of households that rely on groundwater for their daily needs to access the groundwater.

This article examines the connection between a failing, selective piped water supply and groundwater use. It describes the political history of Jakarta's piped water network and the reasons consumers do not, or are unable to, connect to the network. It outlines the link between this failure to connect and the increasingly excessive and unlicensed groundwater use in Jakarta, and its effect on groundwater levels and Jakarta's natural environment. It discusses and assesses current legal and policy reform to address this impending environmental disaster, and asks whether the government is committed to providing a reliable, commercially viable alternative water source to groundwater.

THE POLITICS OF PIPED WATER AND HISTORY OF GROUNDWATER USE

The city of Jakarta is criss-crossed by thirteen rivers and flood channels and has a rich supply of groundwater. Its residents (now estimated to be 12 million) have for centuries relied on groundwater as an important source of water for household, business and industrial purposes. However, historically the importance of groundwater has not been reflected in law or government policy. Nor has effective groundwater management been a high priority. The emphasis of groundwater regulation has been on exploitation rather than sustainable use, and little value has been attributed to groundwater. Its economic, environmental, social and cultural value has not been adequately recognised by the government or by consumers.

Instead, successive governments have focussed on piped water, although adequate universal coverage has not been a priority. Rather, levels of access and quality of piped water have reflected the politics of the various administrations. This has influenced consumer choices and preference for water sources, and has in fact encouraged the use of alternative sources and distribution methods, in particular groundwater.

Prior to the construction of Jakarta's first piped water network, Jakarta's population relied on different combinations of river water and shallow groundwater for its water supply. However in 1870, the Dutch colonial government began constructing an artesian water network intended for use by European residents living in the small 'developed' central area of the city. As a result of this exclusion, the local population continued to rely on surface waters for all of their water supply and sanitation needs as it was the most affordable and convenient water source. Artesian water supply was only made available to local populations at the turn of the century when it was extended minimally to local residents through artesian hydrants.[1]

Following considerable debate over the need for increased water production from a more favourable source, in the 1920s the Dutch government developed a spring water piped network, which expanded on the existing infrastructure. Again the supply was intended to predominantly serve the European residents of Jakarta and was seen as essential based on scientific discoveries of the connection between drinking water consumption and health. European residents with household connections paid half the price for water than did local residents who used public hydrants and so Europeans used most of the water supplied through the network. Local households again continued to rely on other cheaper water sources for economic reasons and convenience.[2] The use of these other sources also prompted the view that local residents were undeveloped as they used traditional sensory assessments to determine water quality without reference to new scientific discoveries.[3] The continued use of non-piped water sources was seen by the colonial government as 'temporary solutions' for the local population who were not yet fully 'modern'.[4]

The piped water network built by the Dutch was badly damaged during the four year battle for independence from 1945–1949, and could not meet the demand for water as Jakarta's population increased.[5] Soekarno, Indonesia's charismatic first president, had a vision for Jakarta as a modern,

---

[1] Four public hydrants were installed for over 116,740 residents. *See* Michelle E. Kooy, Relations of Power, Networks of Water: Governing Urban Waters, Spaces, and Populations in (Post)Colonial Jakarta 53 (Unpublished PhD thesis, University of British Columbia, 2008).

[2] *Id.* at 55, 56, 142.

[3] Sensory assessment of water is still used to determine water quality. For example, a taxi driver interviewed on 28 October 2008 stated that to test the quality of water, he makes a cup of tea and examines the colour of the tea.

[4] *See* Kooy, note 1 above at 83.

[5] Jakarta's population increased dramatically during this time, rising from 8,00,000 in 1945 to 3.15 million in 1965. *See* S. Soetrisno, 'Groundwater Management Problem: Comparative City Case Studies of Jakarta and Bandung, Indonesia', *in* John Chilton ed., *Groundwater in the Urban Environment- Selected City Profiles* 63 (Rotterdam: Balkerna, 1999) and Demographia, World Metropolitan Areas Over 50,00,000: Population Change from 1965: Total, Core & Suburbs, available at http://www.demographia.com/db-worldmetro5m-1965.htm.

progressive, independent city which would be demonstrated by grand public monuments and highly visible infrastructure projects. The new government began rehabilitating existing water network pipes in elite residential areas and constructing water treatment plants. With superior modern water treatment technology, Jakarta could utilise surface water rather than rely on artesian or distant spring water.[6]

Two new water treatment plants, completed in 1957 and 1966 respectively, increased the volume of water available to Jakartan residents by almost ten times the amount circulating during the colonial era.[7] However, there was little increase in the number of residents who had access to the piped water, but rather an increase in the volume of water that a minority of residents could access. During this period, no money was invested in improving access for informally settled low-income areas, not even through basic infrastructure such as public hydrants. By the end of the 1950s, around 80 per cent of the population continued to rely on sources other than piped water, such as increasingly polluted groundwater, rivers and canals.

During the Soekarno era, the legal approach to water was reflected in the 1945 Constitution, which stated that water was to be controlled by the State and used for the greatest benefit of the people.[8] However in practice, politics guided the distribution and expansion of the piped water network to the elite. Groundwater was not a government priority.

From 1965 to the late 1980s, following the violent transition from Soekarno's Old Order government to Soeharto's New Order, the government expanded the network to assist and sustain industrial economic growth and to supply upper class residential areas that were politically supportive of the New Order.[9] This was in line with Soeharto's vision for Indonesia of economic development and social stability. During this period, improvements focused on the expansion of the network with a priority of construction, not rehabilitation; the construction of two additional large scale water treatment plants; and new investment into small-scale treatment plants. Water supply production capacity increased threefold, however distribution of the water through the piped network was extended to less than one-quarter of the city's population, and covered less than half of the urban area.

In spite of Soeharto's vision favouring industry and elite residential areas, in 1995 more than 70 per cent of industries in Jakarta used groundwater either because piped water supplies were inadequate or because it was substantially cheaper to do so. For the same reason, many upper class residential areas also did not connect to the network or use piped water in spite of being connected to the network. PAM Jaya, Jakarta's regional water supply company, acknowledged at the time that it could supply barely 60 per cent of the city's daily demand of 1.7 million cubic metres of water.[10]

Little was done during the Soeharto era to service the urban poor in spite of unprecedented sustained economic growth rates that averaged nearly 6 per cent per year from 1975 to 1996.[11] Those without access to piped water (nearly 80 per cent of the population) then either built their own shallow groundwater wells or connected to other household's wells, or paid up to 20 per cent of their monthly income for water from public hydrants.[12] Sewerage systems were non-existent, a legacy of a government policy treating sewage as a 'private concern'.[13]

In 1990 it was estimated that the amount of piped water consumed by residents in Jakarta was 128 million cubic metres per year. However, the un-served and underserved residents of Jakarta abstracted almost twice this amount, between 200–250 million cubic metres per year of groundwater.[14]

---

[6] The city's new water treatment plants changed the bulk of the source of water supply from mountain spring water to treated surface water taken from the city's flood canals.

[7] *See* Kooy, note 1 above at 62.

[8] Indonesian Constitution 1945, Article 33(3).

[9] *See* Kooy, note 1 above at 68, 70 and 149.

[10] Michael Richardson, 'Water Woes Threaten Growth in Indonesia', *New York Times*, 26 June 1995, available at http://www.iht.com/articles/1995/06/26/water.php.

[11] Paul McCarthy, Understanding Slums: Case Studies for the Global Report on Human Settlements 2003. The Case of Jakarta, Indonesia, World Bank, 2 (2003) available at www.ucl.ac.uk/dpu-projects/Global_Report/pdfs/ Jakarta.pdf.

[12] Karen Bakker et al., Disconnected: Poverty, Water Supply And Development In Jakarta 11 (United Nations Development Programme, Human Development Report Office Occasional Paper, 2006), available at http:// hdr.undp.org/en/reports/global/hdr2006/papers/bakker_et_al1.pdf.

[13] *Id.* at 11. Less than two per cent of households in Jakarta are connected to a sewerage system and the majority of wastewater is disposed directly to rivers, canals, or to often poorly functioning septic tanks. *Id.* at 14.

[14] *See* Kooy, note 1 above at 90–91.

By the 1980s, concerns were growing over water quality in shallow groundwater. The lack of an effective sewerage and waste collection system was contaminating groundwater, in certain areas groundwater depletion and/or salinisation had occurred, and reported cases of water-borne diseases rose.[15] Decades of uncontrolled groundwater abstraction had also changed the hydraulic situation in Jakarta from a relatively undisturbed stage to a stage of exhaustion.[16]

During this time, water laws cemented and elaborated on the basic principles set out in the Constitution. The government passed Law No.11/1974 on Water Resources Development[17] which expanded on the Constitution by stating that the use of water from its source for household purposes is free of charge and has no licence requirements. This principle was further elaborated in Government Regulation No.22/1982 on Water Management which set out that the use of water for public drinking water is a primary priority; that everyone has the right to use water as a primary necessity of life; that a licence is not required for this water usage, nor does it have to be paid for.[18] In spite of these requirements in law in relation to the use of water from its source, in practice the piped water network was developed and expanded to support industrial growth and to supply upper class residential areas that were politically supportive of the New Order government.

Groundwater management was outside the scope of Law No.11/1974, which was limited to the quantity management of surface water. The responsibility for groundwater management lay with the Department of Energy and Mineral Resources as groundwater was seen mainly as a mining resource and not as a renewable resource. Regulation of groundwater consisted of guidelines with an emphasis on the technical aspects of groundwater extraction. The historical emphasis of groundwater regulation was therefore on exploitation rather than sustainable use.

As both the piped and non-piped water quality in Jakarta continued to decline, the city's water supply increasingly created serious problems for human health and the environment, and hindered economic development and poverty alleviation. Private sector participation was seen as the best solution to address the decline in Jakarta's piped and non-piped water and ultimately to assist in the transition from a predominant use of groundwater to piped water.

WHY HOPES FOR A PRIVATISATION-LED SHIFT TO PIPED WATER FAILED

Prior to privatisation, Jakarta's centralised piped water network had numerous problems. These included a water supply distribution network that was not capable of distributing the available water; limited water coverage; poor water quality; high levels of unaccounted-for-water; inequitable access to piped water; and a regional water supply company that was inefficient and underperforming.

In January 1998, two 25 year contracts for the management and expansion of Jakarta's water supply system were awarded to Thames Water International (United Kingdom) and Ondeo (Suez-Lyonnaise des Eaux) (France). Jakarta was split in half, using the Ciliwung River as the dividing line and Thames was awarded the eastern half of Jakarta and Ondeo the western half. The contracts were signed with PAM Jaya (the Jakarta regional water supply company) which retained ownership of the water supply assets.[19] The two companies promised to improve water production, reduce water losses, increase the number of connections, improve service coverage ratio, increase the volume of water billed and improve the service standards. PAM Jaya agreed to assist the two companies to force the closure of deep wells[20] where piped water was available.[21] The two private water companies also

---

[15] See Bakker et al., note 12 above at 11.

[16] Urban Groundwater Database (Information Supplied by Haryadi Tirtomihardjo, Directorate of Environmental Geology, Bandung, Indonesia, 2 July 1996), available at http:// www. scar.utoronto.ca/~gwater/IAHCGUA/UGD/ jakarta.html.

[17] Law No.11/1974 replaced the Dutch General Water Regulations of 1936. Law No.11/1974 contains 17 Articles and is a framework Act with further elaboration in implementing regulations.

[18] See Articles 13-29 of Government Regulation No.22/1982.

[19] The contracts were renegotiated in 2001 following the devastating Asian financial crisis of 1998.

[20] Deep wells are defined in the Contract as those wells that require groundwater permits under the provisions of DKI Jakarta Regulation No.10 of 1998, as amended and replaced (see Article 1.1 of the Contract).

[21] See Articles 9.2(b) and 12 of the Contract.

signed a Memorandum of Understanding in 2004 that required them to supply clean water in order to minimise groundwater use.[22]

However, according to Achmad Lanti, the former Chair of the Jakarta Water Supply Regulatory Body,[23] the private companies are as inefficient as the previous public utility they replaced. Their only incentive is to improve the bill collection system and shut down illegal private wells. The companies have no incentive to improve the system as a whole and have not met key performance targets set out in their contracts.[24] In spite of this underperformance, water tariffs in Jakarta are reportedly the highest in Indonesia. This bleak picture impacts on the quality, quantity and continuity of Jakarta's piped water, and on the willingness of consumers to connect to the network.

Most households base their choice of water source and distribution methods on the least cost solution that meets their basic needs. If piped water cannot meet their basic needs, households will not connect (or advocate for connection) to the network, or they will not utilise the network even though it is connected to their residence.

In 2005, official estimates of the number of households connected to the central network ranged from 46 per cent to 56 per cent.[25] However, if informal (and therefore illegal) settlements are included, only an estimated 25 per cent of Jakarta's 12 million habitants are actually connected.[26] These figures also do not differentiate between households that want access to the piped water and those that have opted out of the network in favour of other alternatives. A significant number of households choose not to connect to the centralised network or are in fact connected but are zero consumption customers.[27] In 2006, there were 110,000 zero consumption households, or around fifteen per cent of network customers. Data from PT PAM Lyonnaise Jaya (Palyja), the private water operator in the western half of Jakarta, shows that in its operating area, 86 per cent of the zero consumption customers 'simply chose to rely on other water sources'[28] such as deep and shallow groundwater wells with filters, pumps, and pipes, and bought bottled water.

There are several reasons why piped water is not the least cost solution that meets the basic needs of households. A high initial expenditure for connection fees and additional charges added to the monthly bill[29] can mean that the total cost of piped water is more expensive than informal water supply methods.[30] The initial connection fee affects the urban poor in particular, with connection fees more than a month's minimum wage, often to be paid in a lump sum. Many poor households have irregular incomes, making this up-front connection fee prohibitive. Connection fees are also more expensive the further the dwelling is from the piped network.[31] Poor households are more likely to be in areas without networks or in areas of lower network density.[32] Methods of payment for monthly bills

---

[22] Mustaqim Adamrah, 'Palyja, Aetra Boost Supply, Reduce Use of Groundwater', *Jakarta Post*, 9 July 2008, available at http://www.thejakartapost.com/news/2008/07/09/palyja-aetra-boost-supply-reduce-use-groundwater.html.

[23] The Jakarta Water Supply Regulatory Body was established to independently monitor and regulate water policies and tariffs at the macro level, supervise the private water companies on behalf of the Governor of Jakarta and the Ministry of Public Works, and mediate disputes between PAM Jaya (the Jakarta regional water supply company) and the two private companies. This was provided for in the revised contracts signed by the private companies in 2001 and subsequently in a regulation issued by the Governor of Jakarta. *See* Schedule 20 of the Contract for a draft of Keputusan Guberur Propinsi Daerah Khusus Ibukota Jakarta Pembentukan Badan Pengatur Pelayanan Air Minum (Decree of the Provincial Governor of the Special Region of the Capital Jakarta regarding the Regulatory Body).

[24] International Consortium of Investigative Journalists, Water and Politics in the Fall of Suharto (February 2003), available at http://www.geocities.com/RainForest/vines/4301/ water04.html.

[25] *See* Bakker et al., note 12 above at 13. However, Indonesia's Report on the Achievement of Millennium Development Goals (83, 2007) states that 63 per cent of Jakarta's population is connected to the piped network. *See* Ministry of National Development Planning, Report on the Achievement of Millennium Development Goals, Indonesia 2007 (Jakarta: Ministry of National Development Planning, 2007), available at http://www.undp.or.id/pubs/docs/ MDG%20Report%202007.pdf.

[26] *See* Bakker et al., note 12 above at 13.

[27] Zero consumption customers are customers who are connected to the piped water network and pay monthly connections fees and other costs, but do not use the piped water.

[28] *See* Kooy, note 1 above at 107.

[29] Monthly bills also include more than charges per unit volumes of water consumed. Additional charges such as the meter fee and a fixed charge are also included in the monthly bill.

[30] *See* Bakker et al., note 12 above at 13.

[31] Connection fees are approximately Rp.10,00,000 (USD 100): interview with KRuHA (People's Coalition for the Right to Water), 11 November 2008.

[32] See Bakker et al., note 12 above at 17.

are also not convenient for low income households, with payment by internet or by automatic teller machine transfer a common method of payment. However, many low income households do not have bank accounts or internet connections in their homes.

Informal settlements face additional legal obstacles to connection. Access to public services in Jakarta is contingent on two requirements. To connect to the network, residents must have state recognised residency for Jakarta and legal occupation of land documented by the State. The former takes the form of a Personal Identification Card which pre-supposes government permission to live and work in Jakarta[33] and the latter means that residents must have a certificate for the land on which they live, indicating a State sanctioned right to live on or use the land. Many poor households do not have such certificates, either because they forgo the formal registration process, are renters of poor quality housing, or are illegal 'squatters' of unoccupied public or privately owned land. In other cases, residents just do not have certificates to prove legal occupation of the land.[34] They are therefore considered to be illegal occupiers of the land, even though many have lived or worked on the land for years, have paid for the land, and have electricity and in some cases telephone connections provided by State owned corporations. The result is that many of the city's lowest income residents do not qualify for public water supply services,[35] and must seek water supply from alternative sources such as groundwater.

Aside from expenses and difficulties in regard to connection, there are also ongoing additional expenses as a result of the low network pressure for all households. Low pressure and intermittent water flow requires financial costs for storage containers and pumps, and physical space to store the containers and pumps.[36] Jakarta is a city with an extremely high population density and many residents do not have the money or the space for such additional requirements.[37] In the area of Pluit and Ancol in North Jakarta, water provided by Palyja only flows between 2 am and 4 am and if residents do not stay awake to collect water to store in the early hours of the morning, they are forced to purchase water from hydrants or water vendors at an increased price. Consumers must still pay Palyja for their piped water connection. This is the case even for wealthy households in the area, who install pumps on the piped network to ensure they are able to access piped water. However, in some cases even they are unable to extract the water from the piped network, but are charged anyway because when the pumps suck air not water, the water meters continue to register water consumption.[38]

Water tariffs are also high, recorded as the highest in Indonesia.[39] Since privatisation in 1998, tariffs have risen dramatically, increasing by fifteen per cent in February 1998, 35 per cent in April 2001,

---

[33] *See* for example Article 56 of Jakarta Government Regulation No.8/2007 on Public Order which states that every person intending to live and settle in Jakarta must have, among other things, '. . . d. skills and qualifications; . . . e. a guaranteed place of abode and guaranteed employment . . .'.

[34] There are over 80 million land parcels in Indonesia, of which only 17 million are currently registered. *See* World Bank, Land Policy, Management and Administration, (Washington: World Bank, 2005); Human Rights Watch, Condemned Communities: Forced Evictions in Jakarta (New York: Human Rights Watch, 2006).

[35] This is contrary to the requirement that no households should be denied the right to water on the grounds of their housing or land status: United Nations Committee on Economic, Social and Cultural Rights, General Comment 15 (The Right to Water), UN Doc. E/C.12/2002/11, para.16(c). Indonesia ratified the International Covenant on Economic, Social and Cultural Rights in 2005 and is discussed further below. It is also contrary to the statement by Indonesia's Constitutional Court that 'a person's need for water does not depend on their residence': Judicial Review of the Water Resources Law (No 7/2004) No 058-059-060-063/PUU-II/2004 and 008/PUU-III/2005, at 488.

[36] *See* Kooy, note 1 above at 174.

[37] In 2003, it was estimated that five per cent of Indonesia's 210 million people lived in Jakarta which is only 0.03 per cent of Indonesia's total land mass. *See* McCarthy, note 11 above at 3.

[38] 'Dari Saluran Tersendat sampai Sambungan Liar' ('From Blocked Pipes to Illegal Connections'), *Kompas*, 9 May 2005, available at http://westjavawater.blogspot.com/2005/05/ dari-saluran-tersendat-sampai.html.

[39] 'Sutiyoso Ngotot Naikkan Tarif Air Minum' ('Sutiyoso Refuses to Increase Water Tariffs'), *Suara Pembaharuan*, 4 February 2006, available at http:// www.suarapembaruan.com/News/2006/02/04/Utama/ ut01.htm. See also Anonymous 'Jakarta, Tarif Air-nya Tertinggi di Kawasan Asia Tenggara' ('Jakarta, Water Tariffs the Highest in Southeast Asia'), 2008, http:// www.lintasberita.com/Bisnis/ Jakarta_Tarif_Air_nya_Tertinggi_di_Kawasan_Asia_Tenggara.

40 per cent in April 2003[40] and 30 per cent in January 2004.[41] In early 2004 it was determined that there would be a regular increase of tariffs every six months until 2007 without the previously required approval by the Jakarta regional parliament.[42] In 2007, the increase was ten per cent, and Palyja has submitted a proposal to the Jakarta administration to increase water tariffs by an average of 22.7 per cent in 2009.[43]

The low quality of piped water is a further disincentive for rich and poor households to connect to the network. Many residents doubt the quality of piped water,[44] seeing no significant difference between the water quality of piped water and groundwater sources (excluding North Jakarta, a poorer, more water deficient area in Jakarta, where the groundwater is brackish and cannot be used for household purposes).

There is good reason for Jakartan residents to be suspicious of the quality of piped water. As a result of delayed network rehabilitation and negative pipe pressure, the vacuum within network pipes allows wastewater (pipes flow through wastewater gutters and storm drains) to be sucked into the pipes, contaminating the water supply. A 2004 research report by the Ministry of Health indicated that only 46 per cent of the piped water samples in Jakarta met its requirements for 'clean water' standard.[45] Residents can also see and smell the poor quality of the water. For example in Penjaringan in North Jakarta, piped water is brown-black in colour and sometimes mixed with dirt and mud.[46]

Many Jakartans are therefore ambivalent about piped water as it is costly, unreliable and does not provide noticeable benefits when compared with groundwater.[47] On the other hand, groundwater is free, excluding the cost of subsidised electricity used to pump the water; the quality is sufficient for most household uses; and Jakartans have relied on groundwater as a source of water for over a century.

However, it is not only households that experience problems with Jakarta's centralised piped water network. As recently as the end of 2008 the government acknowledged that 'it cannot provide surface water' adequately,[48] and is only capable of supplying 54 per cent of the water needs

---

[40] PT Thames Pam Jaya (at the time majority owned by Thames Water International) and Palyja (majority owned by Suez) threatened to break their contracts if the price increase was rejected. If they pulled out, PAM Jaya would have been liable for a Rp.3 trillion (USD 362 million) penalty fee under the contracts. The fee was to reimburse the companies' costs and losses, as well as to pay 50 per cent of projected profits for the remainder of the contractual term. *See* Bill Guerin, 'Indonesia: How Not to Privatize Water', *Asia Times*, 19 November 2003, available at http:// www.greatlakesdirectory.org/ 111903_great_lakes_privatization.htm and Bambang Nurbianto, 'PAM Jaya Could Face Rp. 3 trillion Fine for Any Contract Termination', *Jakarta Post*, 17 November 2003, available at http://www .thejakartapost.com/news/2003/ 11/17/pam-jaya-could-face-rp-3t-fine-any-contract-termination.html?1#1.

[41] Damar Harsanto, 'City Administration Quietly Raises Water Rates by 30 Percent', *Jakarta Post*, 6 January 2004, available at http://www.thejakartapost.com/news/2004/01/06/city-administration-quietly-raises-water-rates-30-percent.html.

[42] Surat DPRD DKI Jakarta Nomor 550/-1.778.1 tanggal 23 Juli 2004 (Letter of the Jakarta Province Regional Parliament Number 550/-1.778.1 dated 23 July 2004).

[43] Anonymous, 'Water Quality Tops Complaints in Consumer Survey', *Jakarta Post*, 22 December 2008, available at http:/ /www.the-jakartapost.com/news/2008/12/22/water-quality-tops-complaints-consumer-survey.html. According to Palyja Commissioner Bernard Lafronge, this is in accordance with the 20 per cent inflation rate of the past two years; Agnes Winarti, 'Palyja Seeks Rate Hike, Vows Investment', *Jakarta Post*, 8 January 2009, available at http:/ /www.thejakartapost.com/news/2009/01/08/palyja-seeks-rate-hike-vows-investment.html.

[44] The Indonesian Consumer Foundation (YLKI) receives 70 complaints from residents about tap water each month: Agnes Winarti, 'JICA Offers Administration System to Test Water Quality', *Jakarta Post*, 22 January 2009, available at http://www.thejakartapost.com/ news/2009/01/22/jica-offers-administration-system-test-water-quality.html.

[45] *See* Kooy, note 1 above at 177-178. The Decree from the Ministry of Health No. 907, 2002 defines drinking water as treated or untreated water that meets health requirements and can be drunk directly. Clean water is defined as water used for daily needs, which meets health requirements and can be drunk after being boiled. *See* Ministry for People's Welfare, Indonesia Progress Report on the Millennium Development Goals 80 (Jakarata: Ministry for People's Welfare, 2004).

[46] 'Kucuran Air Hanya Bisa Dinikmati Pukul 24.00, PAM Kecewakan Warga Jakarta Utara' ('Water Flow Can Only be Enjoyed at Midnight, PAM Disappoints North Jakarta Residents'), *Harian Terbit*, 24 May 2005, available at http:// westjavawater.blogspot. com/2005/05/kucuran-air-hanya-bisa-dinikmati-mulai.html.

[47] Interview with taxi driver, 12 November 2008.

[48] Acknowledgement of PAM Jaya, which is the government party to the water supply contracts, reported on TvOne in its news story *Krisis Air Tanah* (Groundwater Crisis) on Apa Kabar Indonesia (What's News Indonesia), 31 October 2008 [hereafter Groundwater Crisis].

of Jakarta.[49] Such problems mean that business and industry also demonstrate a strong preference for groundwater and drill deep wells to access a cleaner, cheaper and more reliable source of water. In recent years, the level of extraction has become excessive and groundwater theft is common.

With households, businesses and industry relying to a large extent on groundwater, concern then turns to what effect this is having on groundwater levels and Jakarta's natural environment. As discussed below, this is having a devastating effect.

THE EFFECT OF EXCESSIVE GROUNDWATER USE ON GROUNDWATER LEVELS AND JAKARTA'S NATURAL ENVIRONMENT

As discussed above, groundwater has always been an important source of water for Jakartans, and about 70 per cent of the population continue to depend on it, while the majority of industries in the Jabotabek area[50] also rely on groundwater for their water.[51] Certain government departments also encourage the use of groundwater as an alternative source of water to piped water. For example, the Jakarta Mining Agency advocates deep wells as solution to the city's lack of centralised network coverage and service quality, and in 2007, it advocated that 20 new wells should be drilled so that groundwater could be taken from depths of at least 250 metres to ensure water security in the dry season. In 2007 the Agency mapped out areas in need of wells, 'prioritizing areas with little or no access to piped water'.[52]

This excessive level of groundwater use is exacerbated by the failure to replenish groundwater at a sufficient rate. In 1995 industries, hotels and private consumers in Jakarta were drawing more than 300 million cubic metres of groundwater a year,[53] which was about three times the rate that the aquifers were being replenished.[54] Since these figures were measured, the problem of nonreplenishment has increased as massive buildings and concrete take over natural drainage sites, green areas and open spaces, and human waste and rubbish clogs waterways.[55] Freshwater floods now surge up from the ground during the rainy season and rainwater flows straight into the sea, making no contribution to groundwater.[56]

However, it is the urban poor that are often blamed for exacerbating flooding and for blocking natural drainage sites, green areas and open spaces in the city, and for dumping rubbish into waterways. As a consequence, the city government has stepped up the pace of forced evictions of 'illegal' settlements and places of business in the name of public order and opening up green areas. Jakarta

---

[49] 'Berlebihan, Eksploitasi Air Tanah' ('Excessive, Exploitation of Groundwater'), *Suara Pembaruan*, 18 November 2008, available at http://digilib-ampl.net/detail/ detail.php?row=&tp=kliping&ktg=airminum&kode=7975.

[50] Jabotabek is the term given to the metropolitan area surrounding Jakarta. The area consists of Jakarta (as a province on its own) and the three surrounding regencies of Bekasi and Bogor in West Java and Tangerang in Banten, including the cities of Bogor, Bekasi, and Tangerang.

[51] Seventy per cent of industry relies of groundwater as its water source. *See* Groundwater Crisis, note 48 above.

[52] Adianto P. Simamora, 'City Water Crisis Could See Deeper Wells Dug', *Jakarta Post*, 18 August 2007, available at http:/ /www .thejakartapost.com/news/2007/08/18/city-water-crisis-could-see-deeper-wells-dug.html.

[53] At this time, 3020 deep private wells were registered in Jakarta, but nearly 88 per cent of the total groundwater abstraction wells were unregistered. *See* Nur Endah Shofiani, Reconstruction of Indonesia's Drinking Water Utilities: Assessment and Stakeholders' Perspectives of Private Sector Participation in the Capital Province of Jakarta 5 (Unpublished Masters thesis, Royal Institute of Technology, Stockholm, 2003).

[54] *See* Richardson, note 10 above.

[55] Jakarta is criss-crossed by thirteen rivers and flood channels that were built by the Dutch colonial government to alleviate flooding. However, the channels are ineffective when it rains as they are clogged by garbage dumped into the channels by the city's residents and there has been no proper maintenance of the channels for 25 years. *See* Geoff Thompson, 'Jakarta in Jeopardy', Australian Broadcasting Corporation, 23 September 2008, available at www.abc.net.au/foreign/ content/2008/s2368261.htm.

[56] Editorial, 'Water Scarcity', *Jakarta Post*, 5 April 2008, available at http://www.thejakartapost.com/news/2008/04/05/ editorial-water-scarcity.html.

Government Regulation No.8/2007 on Public Order permits the forced eviction[57] of persons who live or construct dwellings or places of business on green areas, parks, public places, and riverbanks among other areas.[58] A recent example of this increased pace of forced eviction is in North Jakarta, where it is estimated that 24,000 families will loose their homes in the near future to make way for green areas designed to absorb rainfall and prevent chronic flooding. However, it has since been determined that the cleared area will be turned into an international sports stadium, two out door soccer fields, a jogging track and urban forest site.[59] No massive buildings (namely luxury hotels or apartment complexes, factories or shopping malls) built on green areas and natural drainage sites have been demolished.

With the increasing extraction of groundwater, water levels have dropped by one to three metres a year during the last ten years and are locally at 20 to 40 metres below mean sea level.[60] In some areas, this is even more severe, for example, the groundwater level in the Mega Kuningan business area in South Jakarta is dropping by five meters per year.[61] The increasing extraction levels also contribute to rising levels of groundwater salinisation caused by seawater intrusion: where groundwater is replenished, it may instead be replenished by salt water intrusion or by polluted surface water which seeps into the soil. The seawater intrusion also damages the piped network through corrosion.

Groundwater abstraction has also been identified as the main cause of observed land subsidence in Jakarta, which has now reached critical levels.[62] As Jakarta develops, notably involving the construction of more luxury hotels, apartments and shopping malls,[63] the city gets heavier, which with the combination of groundwater extraction and the creation of vacuums in the aquifer, pushes the city downwards.[64] Many of these constructions, in particular shopping malls, are reported to grossly exceed their building floor coefficients, meaning they are too big and therefore too heavy for the land space on which they are built.[65]

High rise, low cost housing built by the government as alternative housing for people who are evicted from their homes has also been identified as a source of increased land subsidence. The apartment buildings are densely populated, with over 5,000 people per hectare, and residents must use groundwater because the tap water supply is insufficient.[66]

---

[57] Forced evictions are defined as 'the permanent or temporary removal against their will of individuals, families and/or communities from the homes and/or land which they occupy, without the provision of, and access to, appropriate forms of legal or other protection'. *See* United Nations Committee on Economic, Social and Cultural Rights, General Comment 7: The Right to Adequate Housing (Art. 11.1 of the Covenant): Forced Evictions, para.3, available at http://www. unhchr.ch/tbs/doc.nsf/(Symbol)/ 482a0aced8049067c12563ed005acf9 e?Opendocument.

[58] *See* Articles 6, 12(c), 13(1), 20 and 36(1) of Jakarta Government Regulation No.8/2007 on Public Order.

[59] 'NGO Urges a Humane Eviction of 24,000 Squatters in North Jakarta', *Jakarta Post*, 19 September 2008 and Tifa Asrianti, 'BMW Park Gets Rp.4b for Design, Assessment', *Jakarta Post*, 2 January 2009, available at http:// www.thejakartapost.com/news/2009/01/02/ bmw-park-gets-rp-4b-design-assessment.html.

[60] *See* Tirtomihardjo, note 16 above.

[61] Mustaqim Adamrah, 'New Groundwater Fees Set for Jakarta', *Jakarta Post*, 28 March 2008, available at http://www.thejakartapost. com/news/2008/03/28/new-groundwater-fees-set-jakarta.html.

[62] *See* Tirtomihardjo, note 16 above.

[63] For example, during the period 2000–2005, the former Governor of Jakarta Sutiyoso issued permits for the construction of shopping malls over an area of more than 3 million square metres of land. This is more than twice the area of shopping centres built in Jakarta from 1962–1997. Sri Palupi, Perda tentang Ketertiban Umum Hasil Revisi Perda 11/1988: Upaya Sistematis Mengusir Orang Miskin dengan Dalih Ketertiban Umum (Regional Government Regulation concerning Public Order the Result of a Review of Regional Government Regulation 11/1998: Systematic Efforts to Drive Out the Poor on the Basis of Public Order) (Paper presented during discussions on the controversy of the Regional Government Regulation on Public Order at the Indonesian National Human Rights Commission, Jakarta, October 2007, page 11).

[64] In addition, trees on the mountains surrounding Jakarta have been removed to make way for holiday villas radically reducing the ability of mountains to absorb water. *See* Thompson, note 55 above.

[65] Examples cited include Kuningan City, Satrio Street; Gandaria City; Ciputra World Satrio Street; and Senayan City in South Jakarta, St Moritz, Puri Indah; Central Park, Tanjung Duren; and Season City, Grogol in West Jakarta, and Kelapa Gading Square in North Jakarta. *See* Agnes Winarti, 'Superblocks Pose Risks, Study Says', *Jakarta Post*, 12 February 2009, http://www.thejakartapost.com/ news/2009/02/12/ superblocks-pose-risks-study-says.html.

[66] 'Proyek Rumah Susun Dinilai Percepat Tanah Jakarta Ambles' ('Highrise Project is Deemed to Accelerate Land Subsistence in Jakarta') *Tempointeraktif*, 3 March 2009.

Land subsidence in Jakarta was first observed in 1926, when a Dutch surveyor conducted measurements from Jatinegara (then on the outskirts of east Jakarta) to Tanjung Priok harbour in the north.[67] No further investigations were made until 1978 when buildings and the Sarinah fly-over bridge at M.H.Thamrin Street in Central Jakarta cracked. Flooding in Jakarta in that period also covered a wider area.[68]

The Jakarta Mining Agency has reported on land subsidence in Jakarta over a twelve year period from 1993 to 2005. For the 60 per cent of land in Jakarta that is above sea level, the rates of subsidence are alarming:[69]

Location	Height above Sea Level in 1993 (in metrs)	Height above Sea Level in 2005 (in metrs)	Land Subsidence (in centimetres)
North Jakarta	2.03	1.46	57
West Jakarta	2.32	2.11	21
East Jakarta	11.62	11.45	17
South Jakarta	28.76	28.46	30
North Jakarta	3.42	2.40	102

As Jakarta slowly sinks, the repercussions are potentially disastrous. Flooding is becoming worse, with February 2008 marking the worst floods for three centuries. Fifty four people were killed and the floods caused nearly USD 1 billion in damages.[70] Perhaps even more catastrophic is the estimation that in less than 20 years time, the sea will permanently flood the first two to four kilometres of the coastal area of Jakarta, rendering almost one third of Jakarta uninhabitable and causing the displacement of millions of people.[71] The date that this will happen has even been predicted: 6 December 2025. The impending disaster has been attributed to a combination of Jakarta's overdevelopment which is compressing the land it is built on, the peak of an 18.6-year astronomical tide cycle and the depletion of groundwater caused by factories, hotels and wealthy residents drilling deep water bores to bypass the piped water network.[72]

The Jakarta administration is beginning to acknowledge the problems caused by excessive groundwater depletion and to take action.[73] However, stopping the overuse of groundwater is not easy, as aquifer systems are large, groundwater pumping is difficult to monitor, users unorganised,

---

[67] P. Suharto, Waterpas Teliti di Indonesia' Risalah Ilmiah ('Level Instruments Precise in Indonesia' Scientific Journals) (Bandung: Geodesi ITB, Bandung, 1971), in Rochman Djaja et al., Land Subsidence of Jakarta Metropolitan Area, Indonesia (3rd FIG Regional Conference Jakarta, Indonesia, October 3–7, 2004).

[68] LM Husoit et al., Pengaruh Alami Geologi dan Berat Bangunan terhadap Penurunan Muka Tanah dan Terjadinya Genangan Air (The Natural Influence of Geology and Weight of Buildings on Land Subsistence and the Occurrence of Pools of Water) (One Day Seminar on Mining Geology, Jakarta Mining Agency, Jakarta, Indonesia, 1997).

[69] See Indonesia: Environment, Science & Technology, and Health Highlights February-March 2005, US Embassy, Jakarta, available at http://www.usembassyjakarta.org/econ/ ESTH_highlight_feb-march05.html.

[70] See Thompson, note 55 above.

[71] By 2025 the population of Jakarta may reach 24.9 million, not counting millions more in surrounding areas: Pepe Escobar 'Book Review. The accumulation of the wretched. Planet of Slums by Mike Davis', Asia Times, 20 May 2009 available at http://www.atimes .com/atimes/Front_Page/ HE20Aa01.html.

[72] Climate change is not a major reason for this permanent flooding. By 2025, estimates from the Intergovernmental Panel on Climate Change (IPCC) show sea levels will have risen by only about five centimetres. See Anonymous, 'Indonesia's Thirsty Capital is a Sinking City', AFP, 15 April 2008, available at http://www.abc.net.au/news/stories/ 2008/04/15/2217414.htm. The findings outlined are the result of a study by the World Bank and Delft Hydrolics.

[73] States should ensure that everyone has a sufficient amount of safe water by combating the depletion of water resources from unsustainable extraction by industry. See United Nations Sub-Commission on the Promotion and Protection of Human Rights, Res. 2006/10, Promotion of the Realization of the Right to Drinking Water and Sanitation, UN Doc.A/HRC/ Sub.1/58/L11 (2006), adopting the Draft Guidelines for the Realization of the Right to Drinking Water and Sanitation, UN Doc.E/CN.4/Sub.2/2005/25 (2005), Section 4.1.

there are hundreds of thousands of consumers, and their numbers increase quickly.[74] The following section outlines the steps the government has actually taken to address the problems and asks whether these steps are sufficient.

REFORM AND ACTION BY THE GOVERNMENT: IS IT ENOUGH?

If the government is to take sufficient steps to address the excessive and unlicensed use of groundwater, it must take a multi-dimensional approach to reform. Not only does the government need to address the current regulation of, and attitude towards, groundwater, it also needs to provide a reliable, commercially viable alternative water source to groundwater to the big extractors. The best way to do this is to improve the piped water supply network, both in regard to the raw water supplied to the network[75] and the capacity of the network to treat and distribute the water efficiently and effectively.

The government has drafted new regulations on water management generally and groundwater particularly, and has started a public awareness campaign on the importance of groundwater. It has also embarked on an ambitious plan with the Asian Development Bank to improve the quality and quantity of raw water supplied to the piped network. However, reform in regard to the network itself remains limited to short term improvements and ancillary measures such as increasing groundwater fees, addressing compliance issues and closing wells and bores. These initiatives are discussed in detail below.

Regulatory Reform

As discussed above, historically the legal framework for the development of water resources management specifically excluded groundwater.[76] However, in 2004 the government passed its framework piece of legislation on water resources management, namely Law No.7/ 2004 on Water Resources. Law No.7/2004 now takes a holistic approach to water management, extending to both surface and groundwater.[77] It also emphasises the need for efficient water use, which is a new concept in Indonesian water management.[78]

Prior to 2004, the responsibility for groundwater management lay with the Department of Energy and Mineral Resources as groundwater was seen mainly as a mining resource and not as a renewable resource. Regulation of groundwater consisted of guidelines with an emphasis on the technical aspects of groundwater extraction. The historical emphasis of groundwater regulation was therefore on exploitation rather than sustainable use, which has contributed to a low awareness of the real value of groundwater in Jakarta. Factory managers for example do not think beyond the fact that unless they sink new and deeper boreholes every two or three years their neighbours will draw away all the water.[79] The low awareness of the value of groundwater can also be seen by the common practice of removing and discarding groundwater when land is excavated for foundations or basements of luxury apartments, hotels, shopping malls and business complexes.[80] In order to encourage users and administrators to place a value on groundwater, and to ensure its sustainable use, it is important that the economic, environmental, social and cultural value of water is recognised. The new Law No.7/2004 recognises this in a very general sense, stating that water has a social, cultural, environmental and

---

[74] O. Braadbaart and F. Braadbaart, 'Policing the Urban Pumping Race: Industrial Groundwater Overexploitation in Indonesia', 25(2) *World Development* 199, 200 (1997).

[75] PAM Jaya, for example, is reportedly reluctant to increase the number of connections because access to raw water is limited (KRuHA interview, 11 November 2008).

[76] *See generally* Law No.11/1974.

[77] Article 1(5) of Law No.7/2004 where the definition of water resource includes both surface and groundwater.

[78] Article 26(6) of Law No.7/2004.

[79] *See* Braadbaart and Braadbaart, note 74 above at 209.

[80] 'Pidana Bagi Penyedot Berlebihan' ('Criminal Sanctions for Excessive Extractors'), *Kompas*, 17 April 2008, available at http://asosiasi .org/tbk/2008/10/pidana-bagi-penyedot-berlebihan/.

economic function.[81] For example, Law No.7/2004 clearly states the need for efficient water use[82] and acknowledges the need to recognise indigenous peoples' rights when exercising control over water resources, although this recognition is heavily qualified.[83] The Law also guarantees everyone's right to obtain water for their minimum daily basic needs, and outlines the government's obligation to ensure that access to water for such purposes is free and is the first priority for water use.[84] However the ability of the State to fulfil this latter obligation is seriously challenged by the combination of the declining water table and salinisation and pollution of groundwater. These factors mean that getting water from shallow wells is becoming more difficult for households that rely on groundwater as their primary source of water.[85] Each year, households must dig deeper wells to access water for household uses[86] or they will not have access to groundwater for basic daily needs.

In this regard, in 2005 Indonesia ratified the International Covenant on Economic, Social and Cultural Rights. Article 11 requires the government to recognise the right of everyone to an adequate standard of living, which includes the right to water. This means that water supply must be sufficient, clean, accessible, affordable and enjoyed without discrimination.[87] The right to water is also guaranteed by Article 28H of the Indonesian Constitution, which was amended in 2000 to include a chapter on human rights. Article 28H states that every person has the right to live in spiritual and physical welfare, has the right to housing, and the right to a good and healthy environment as well as the right to obtain health care.[88] The Indonesian Constitutional Court has determined that Law No.7/2004 is in accordance with the right to water as set out in the Constitution.[89]

The principles in Law No.7/2004 are quite general as this Law is a framework law, the details of which are set out in implementing regulations. The implementing regulation on groundwater, Government Regulation No.43/2008 on Groundwater, was passed in May 2008. The Regulation elaborates on the framework provided in Law No.7/2004 and is an important step towards strategic groundwater management. As set out in the framework Law No.7/2004, its emphasis is no longer just on exploitation, but also on conservation and the sustainable use of groundwater.[90] The Regulation favours the basin approach to groundwater management,[91] whereas in the past, the focus was on the management of boreholes. The management of groundwater as a hydro-geological unit is now central.[92] It appears that the Department of Energy and Mineral Resources is still primarily responsible for groundwater management, although this is not clearly stated.[93]

Importantly, Government Regulation No.43/2008 recognises the connection between groundwater management and surface water management and that the two should be managed in an integrated manner.[94] However, it does not express the need for administrative coordination between the different

---

[81] *See* Article 4 of Law No.7/2004.

[82] Law No.7/2004 clearly states that every person has the responsibility to use water in a conservative manner as possible (Article 26(6)), although it does not state what this means exactly or how it will be enforced.

[83] Article 6(2) and (3) of Law No.7/2004. The qualifications are that the indigenous peoples' rights must not be contrary to the national interest and laws, and the indigenous peoples' rights must still exist and have been consolidated by local government regulation.

[84] *See* Articles 5, 29(3) and 80 of Law No.7/2004 on Water Resources.

[85] *See* Djaja et al., note 67 above.

[86] Interview with Jakarta resident, 13 November 2008.

[87] Sufficient means that water must be sufficient and continuous for personal and domestic uses; clean means safe water that in particular is free from hazardous substances that could endanger human health, and whose colour, odour and taste are acceptable to users); accessible means that water services and facilities are accessible within, or in the immediate vicinity, of each household, educational institution and workplace; affordable means that water can be secured without reducing a person's capacity to acquire other essential goods and services, including food, housing, health services and education..*See* United Nations Committee on Economic, Social and Cultural Rights, note 35 above, Paras. 2 and 12.

[88] Judicial Review of the Water Resources Law (No 7/2004) No 058-059-060-063/PUU-II/2004 and 008/PUU-III/ 2005, at 488.

[89] *Id*. at 485, 491–2.

[90] *See* for example, Elucidation I. General Point 4 and 5 of Government Regulation No.43/2008 on Groundwater.

[91] *See* Elucidation I. General, Point 2; Article 4 of Government Regulation No.43/2008 on Groundwater.

[92] H.H.A. Teeuwen, Groundwater Management and the New Indonesian Water Legislation 8 (Ministry for Transport, Public Works and Water Management, Rijkswaterstaat, The Netherlands, September 2005).

[93] The Department of Energy and Mineral Resources was responsible for drafting the Regulation.

[94] *See* Article 3(1) of Government Regulation No.43/2008 on Groundwater.

government agencies responsible for surface water and groundwater management.[95] The recognition is also limited in very general terms to conservation, efficient utilisation and control of the potential damage to groundwater.[96] This means that plans for groundwater management must prioritise the use of surface water in the relevant river basin;[97] the conservation of groundwater must involve using groundwater as the last alternative;[98] and plans for provision of groundwater must be compiled bearing in mind plans for the provision of surface water.[99]

The general manner in which the connection between groundwater and surface water management is recognised in the Regulation makes this aspect of the Regulation difficult to apply in practice, and the supervision of its implementation even harder. The Regulation does not take into account situations where surface water is inadequate, for example, where the surface water is not an improved water source,[100] is not affordable, the supply is not continuous or where it takes too long to collect the water.[101] It also does not recognise in detail the physical connection between groundwater and surface water.

In regard to efficient use of groundwater, the Regulation recognises the current unsustainable levels of groundwater use. However, several of the requirements to ensure the sustainable use of groundwater are impractical. For example, to prevent the pollution of groundwater, groundwater users must close (and therefore no longer use) wells where the water is polluted.[102] In Jakarta more than half of the population rely on groundwater from wells as a source of water, but more than 60 per cent of wells sampled in 2004 showed e-coli contamination in excess of the regulated drinking water level, despite the majority of these being classified as protected wells. In 2006, the Jakarta Environmental Monitoring Agency (BPHLD Jakarta) estimated that 80 per cent of deep wells were contaminated with e-coli,[103] and in September 2008 concluded that in some areas (West Jakarta and Central Jakarta) contamination was as high as 93 per cent.[104] It is also unclear how this requirement to close polluted wells can be applied consistently with the requirement that the supply of groundwater for daily needs must be prioritised above all other needs,[105] and that the State must 'guarantee everyone's right to obtain water for their minimum daily basic needs'.[106]

The Regulation also notes the occurrence of sea water intrusion into groundwater and land subsidence, stating that the way to avoid this is by prohibiting the extraction of groundwater in coastal areas,[107] and stopping the extraction of groundwater in critical zones and damaged zones.[108] How this will be achieved in practice is unclear, and it will not lead to the desired results without similar commitment to reducing Jakarta's overdevelopment on natural drainage sites, green areas and open spaces. Preventing land subsidence can also be through the creation of artificial recharge areas.[109] All three methods will be further regulated by Ministerial Regulation.[110]

---

[95] *See* Elucidation I. General, Point 1 of Government Regulation No.43/2008 on Groundwater.

[96] Elucidation of Article 3 of Government Regulation No.43/ 2008 on Groundwater.

[97] Article 25(2)(a) of Government Regulation No.43/2008 on Groundwater. This reflects Article 26(6) of the framework law.

[98] Article 42(1)(d) of Government Regulation No.43/2008 on Groundwater.

[99] Article 51(1) of Government Regulation No.43/2008 on Groundwater.

[100] An improved water source is one that is likely to be safe, and includes household connections, public standpipes, boreholes, protected dug wells, protected springs and rainwater collection. *See* WHO and UNICEF, Joint Monitoring Program for Water Supply and Sanitation: Definitions, available at http://www.wssinfo.org/en/ 122_definitions.html.

[101] These are elements of the right to water, as set out in General Comment No.15. *See* United Nations Committee on Economic, Social and Cultural Rights, note 35 above.

[102] Article 46 of Government Regulation No.43/2008 on Groundwater.

[103] World Bank, Economic Impacts of Sanitation in Indonesia: A Five-country Study Conducted in Cambodia, Indonesia Lao PDR, the Philippines, and Vietnam under the Economics of Sanitation Initiative (ESI) 32 (Jakarta: World Bank, Research Report, 2008).

[104] 'Air Tanah Jakarta Semakin Tercemar Bakteri E-coli' ('Jakarta Groundwater Increasingly Contaminated by the E-coli Bacteria'), *Tempointeraktif*, 27 February 2009, available at http://www.tempointeraktif.com/hg/layanan_publik/ 2009/02/27/brk,20090227-162362,id.html.

[105] Article 50(3) of Government Regulation No.43/2008 on Groundwater.

[106] Article 5 of Law No.7/2004.

[107] Article 62(2) of Government Regulation No.43/2008 on Groundwater.

[108] *Id*. Article 63(2).

[109] *Id*. Article 63(3).

[110] *Id*. Article 64.

As mentioned above, the Regulation prioritises the use of groundwater for domestic use (basic daily needs), for small-scale farming in an irrigation system, and as a source of drinking water.[111] The use of groundwater for commercial purposes requires a permit and a fee may be charged.[112] Where licences are required, the Regulation does not set out a clear procedure in relation to licensing and this may discourage users from applying for a licence, and therefore unintentionally encourage unlicensed groundwater use. For example, the Regulation does not stipulate when a decision must be made regarding the application, and has no information about the nature of permit conditions to be set, about amendment to and revocation of the permit, or about the handling of any appeals to be lodged against the permit decision.[113] It merely states that licensing will be further regulated by Ministerial Regulation.[114]

It is questionable whether the financing structure in the Regulation offers sufficient incentives for reducing groundwater usage. A fee instrument is likely to be inappropriate for such a purpose given that the way the fee is determined allocates no value to the groundwater itself.[115] If as stated in the Regulation one of the goals is limiting groundwater use, the Regulation should focus more on the strict application of the permit instrument.[116] In this regard, clear policy frameworks should set out a strategic plan for achieving the reduction of groundwater, especially by industry, luxury hotels and apartments and shopping malls.

The government has also prepared a draft regulation on water usage rights.[117] This draft regulation also contains provisions on licensing. However, it appears that the licensing systems in the Regulation on Groundwater and in the draft regulation on water usage rights differ, although both may cover groundwater. For example, in the Regulation on Groundwater, water usage rights to utilise groundwater for basic daily needs requires a permit if the diameter of the bore is more than five centimetres, the water is not extracted by hand or the water used is more than 100 m^3 per month for each family and is a central distribution point.[118] On the other hand, the draft regulation on water usage rights states that a licence is required for basic daily needs if the manner water is used changes the natural condition of water sources, if the water is used in large amounts (it is unclear what this means), or if it is used for community farming outside the existing irrigation system.[119]

Where no licence is required under the draft regulation on water usage rights, such rights are realised through a water allocation certificate, published on the basis of an inventory of population number, estimated water needs to fulfil basic daily needs and estimates of water needs for community farming in pre-existing systems.[120] The certificate is intended to guarantee the right of the certificate holder to manage the water resource.[121] It is doubtful that households considered to be illegally occupying the land will be issued with such a certificate and therefore they will have no guarantee of free groundwater for personal daily use under the draft regulation. The Regulation on Groundwater makes no provision for such certification.

Another area in which the two regulations differ although both can apply to groundwater, is in relation to the priority of water use. While the draft regulation on water use rights prioritises water use to fulfil daily basic needs, prioritisation between other uses of water is absent. The Regulation on Groundwater clearly sets out the priority of water use.

Finally, an important point of difference between the two regulations is the need for efficient groundwater use. While the Regulation on Groundwater clearly recognises the need for efficient use of groundwater (both shallow and deep),[122] the draft regulation on water usage rights does not contain

---

[111] *Id.* Article 47(1).
[112] *Id.* Article 58.
[113] *See* Teeuwen, note 92 above at 20.
[114] Article 69 of Government Regulation No.43/2008 on Groundwater.
[115] *Id.* Articles 83–84.
[116] *See* Teeuwen, note 92 above at 15.
[117] *See* Draft Regulation on Water Usage Rights, 10 June 2008.
[118] Article 55 of Government Regulation No.43/2008 on Groundwater.
[119] Article 8(1) and (2) of the Draft Regulation on Water Usage Rights.
[120] *Id.* Article 20 (1) and (2).
[121] *Id.* Article 20(6).
[122] *See* Articles 41 and 42 of Government Regulation No.43/2008 on Groundwater.

any such obligations where groundwater is used for fulfilling daily basic needs or for community based agriculture. Obligations regarding water use are only imposed on those holding a licence, which neither of the users of groundwater for daily basic needs or for community agriculture are required to have. However, Law No.7/2004, on which this draft regulation is based, has an overarching requirement for efficient groundwater use.

Under the draft regulation, holders of water usage rights are obliged to give the local community access to the water resources over which they hold rights, for the purpose of fulfilling the community's basic daily needs.[123] In this respect it is important the local community is obliged to use the water efficiently so as to avoid any conflict over water usage.

Public Awareness Campaigns

In addition to the regulatory reform outlined above, the Jakarta administration has begun a public awareness campaign regarding the importance of groundwater in the form of television advertising addressing groundwater conservation.[124] The slogan of the campaign is 'selamatkan air tanah untuk tanah air' (save groundwater for the sake of our nation).[125] The Jakarta Mining Agency is also promoting the 5R Programme (reduce, reuse, recycle, recharge and recover) to industries and companies, which is aimed at conserving groundwater.[126]

In this regard, the government has also introduced a requirement that homeowners and commercial building owners build facilities to store rainwater and to channel it underground. This may be either in the form of rainwater run-off pools, catchment basins or biopore cylinders.[127]

It is intended that the facilities will prevent repeat flooding, water scarcity and will recharge groundwater in the event of drought.[128] The target in regard to biopore cylinders for 2009 is to create five million in five municipalities across Jakarta.[129] However, it is unclear how this will be implemented in practice given the high population density of Jakarta. It is also unclear how its implementation will be supervised given Jakarta's official population is around 8.5 million residents[130] and a biopore cylinder unit can be as small as ten centimetres in diameter for a house on land measuring up to 21 square metres. The cost may also be prohibitive for many low income residents, with a house on land this size requiring at least three cylinders at a cost of over half a month's official minimum wage.

The Need for Adequate Surface Water

While regulatory reform and public awareness campaigns are important steps in addressing excessive groundwater depletion, as mentioned above, the strategy must also include providing a reliable, commercially viable alternative water source to groundwater to the big extractors. The best way to do this

---

[123] *See* Article 35 of the Draft Regulation on Water Usage Rights.

[124] The advertisements observed were shown on TvOne in October 2008.

[125] The Sub-Commission Guidelines state that 'States should adopt measures to prevent over-consumption and promote efficient water use, such as public education, dissemination of appropriate conservation technologies, and, as necessary, restrictions on water use beyond an acceptable consumption threshold, including through the imposition of charges'. *See* United Nations Sub-Commission on the Promotion and Protection of Human Rights, note 73 above, Section 4.2.

[126] Anonymous, '1,000 Firms Fined for Overusing Groundwater', *Jakarta Post*, 12 December 2008, available at http://www.thejakartapost .com/news/2008/12/12/1000-firms-fined-overusing-groundwater.html.

[127] 'Biopores are tunnels bored into the soil that enable organisms to become more active and plants to take root more easily. Such processes create hollow spaces inside the soil that are filled with air, and these air-filled spaces function as channels to absorb water more readily'. Biopores allow the soil to absorb water and minimise the possibility of water inundating soil's surface. This in turn reduces flooding since the water is directly absorbed into the soil. *See* Anonymous, 'Biopore Infiltration Holes: A Flood Prevention Method', *Jakarta Post*, 26 June 2008, available at http://www.nowpublic.com/environment/biopore-infiltration-holes-flood-prevention-method-0.

[128] Adianto P. Simamora, 'Households May Have to Store Rainwater', *Jakarta Post*, 9 February 2009, available at http://www.thejakarta post.com/news/2009/02/09/households-may-have-store-rainwater.html.

[129] Triwik Kurniasari, 'City Aims to Build 5m Biopores this Year', *Jakarta Post*, 1 April 2009, available at http://www.thejakartapost.com/ news/2009/04/01/city-aims-build-5m-biopores-year.html. Jakarta currently has only 335,590 biopores, far below the recommended 76 million.

[130] Dinas Kependudukan dan Catatan Sipil (Population and Civil Registry Office), December 2008, available at http://www.kependudu kancapil.go.id /index.php?Itemid=63&id=4&option=com_content&view=article.

is to improve the piped water supply network, both in regard to the raw water supplied to the network and the capacity of the network to treat and distribute the water efficiently and effectively.

Quality of Raw Water Supplied to the Network

In regard to the first factor, namely the quality of raw water supplied to the network, 80 per cent of Jakarta's water supply comes from the Citarum River Basin in West Java and flows through the West Tarum Canal, a 68 kilometre long canal that was opened in 1968. The Citarum River was named one of the most highly polluted rivers in Indonesia in a 2007 study by the Indonesian Department for the Environment. It has a biochemical oxygen demand (BOD) of 155.[131] The Asian Development Bank has even called the Citarum River the world's dirtiest river.[132] It is polluted by industrial waste, household waste and garbage from the settlements along its banks, agricultural waste and silt from increasing erosion and landslides in the river basin area. As a result, the quality of the raw water treated by the private piped water companies is poor and requires extensive treatment, the subsequent effect of which is that piped water often has a strong smell of chlorine and other chemicals and does not run clear. Many households and businesses do not want to use the piped water as it smells and is discoloured.[133]

In an effort to address this problem, in December 2008, the Indonesian government and the Asian Development Bank signed a USD 500 million loan for the Integrated Citarum Water Resources Management Investment Program (ICWRMIP). Through the ICWRMIP, the Bank will support the rehabilitation of the West Tarum Canal. The main objective of the rehabilitation work is to improve the flow and quality of water that supplies Jakarta,[134] in addition to supplying the water requirements of industrial establishments and about 52,800 hectares of farmland.[135] At present the Canal operates at 70 per cent capacity only and leakage from the Canal is more than ten cubic metres per second.[136] However, the ICWRMIP is not without controversy. It is feared that the rehabilitation of the Canal will result in at least 872 households being evicted from the Citarum River Basin without an appropriate resettlement plan.[137] The families will not be compensated for lost land as the land is considered 'State land'.[138] The rehabilitation project may also cause temporary problems for Jakarta's piped water supply as canal dredging will disperse bottom sediments.

Capacity of the Network to Treat and Distribute Water

In regard to the second factor, namely the capacity of the network to treat and distribute the water efficiently and effectively, the government has only recently begun to publicly acknowledge the connection between a poor water distribution network and excessive groundwater use.[139] Limited initiatives have

---

[131] BOD is the total oxygen required by micro-organisms to decompose organic substances in sewerage. *See* Adianto P. Simamora, 'Most Rivers Face Severe Pollution, Study Finds', *Jakarta Post*, 2 December 2008, available at http:// www.thejakartapost.com/news/2008/12/02/ most-rivers-face-severe-pollution-study-finds.html. Moderately polluted rivers may have a BOD value in the range of two to eight mg/L.

[132] Anonymous, 'ADB Lends $500 mln to Clean Up Dirtiest River', *Jakarta Post*, 5 December 2008, available at http:// www.thejakarta post.com/news/2008/12/05/adb-lends-500-mln-clean-dirtiest-river.html.

[133] In an interview with a taxi diver in October 2008, the driver commented that his wife did not want to use piped water as it had been treated with chemicals which 'could not be good for you'. It could also be seen in a comment by a resident of the Semanan subdistrict in West Jakarta that '[groundwater] is much better than the pricey tap water which has been mixed with chemical substances'. *See* Bambang Nurbianto, 'Poor People Pay More for Water', *Jakarta Post*, 11 July 2005, available at http:// westjavawater.blogspot .com/2005/07/poor-people-pay-more-for-water.html.

[134] The project is to supply water to 200,000 more households in Jakarta, as it will ultimately increase Jakarta's water supply by 2.5 per cent annually. *See* Anonymous, note 132 above.

[135] Directorate General of Water Resources, Indonesia: Integrated Citarum Water Resources Management Investment Program (Resettlement Plan, Document Stage: Draft for Consultation, Project Number: 37049-01-03, ii, 6-7, August 2008).

[136] Directorate General of Water Resources, Public Information Booklet: Integrated Citarum Water Resources Management Investment Program, available at http:// www.adb.org/Documents/Resettlement_Plans/INO/ 37049/ 37049-06-INO-RP.pdf. [hereafter Public Information Booklet].

[137] Yuli Tri Suwarni, 'ADB Urged to Withdraw Investment in Citarum', *Jakarta Post*, 5 December 2008, available at http:/ /www.thejakarta post.com/news/2008/12/05/adb-urged-withdraw-investment-citarum.html.

[138] *See* Directorate General of Water Resources, note 136 above.

[139] Triwik Kurniasari, 'City to Propose Draft on New Groundwater Tariff', *Jakarta Post*, 4 March 2009, available at http://www.thejakarta post.com/news/2009/03/04/ city-propose-new-groundwater-tariff.html.

been introduced in this respect, all of which are short term and not directed at sustainable improve-ments to the piped water network. The initiatives are directed towards increasing groundwater fees, closing wells and bores and compliance issues.

The Jakarta administration plans to increase current groundwater fees by up to sixteen times their current rate in an effort to balance the prices set by the piped water network providers, and to reduce groundwater consumption and land subsidence.[140] Fees for groundwater have always been far below those for piped water, for example, piped water ranges from Rp.1,050 to Rp.12,550 per cubic metre, whereas groundwater is only Rp.525 per cubic metre for luxury homes and up to Rp.3,300 for industry.[141] Groundwater that does not require a licence and that is used for basic daily needs is free. The proposed new fees for groundwater would be on par or higher than piped water, ranging from Rp.8,800 (luxury homes) to Rp.23,000 (industry) per cubic metre. The government has stated that the components for the proposed new fees are based on costs such as replacing the groundwater taken, with the remainder being environmental compensation costs.[142] These components will ideally increase awareness of the real value of groundwater as well as the need for its efficient and sustainable use. However, there is no indication that the increased revenue from groundwater will be ring-fenced and utilised to promote the sustainable use of groundwater in practice. Rather it is likely that the additional money will be used for general purposes as part of the wider government budget.

The Jakarta government has also begun to actively implement the requirement in the contracts with the private operators that it assist in closing deep wells where piped water is available.[143] Re-cently, commercial jeans-laundry plants were given a deadline of early December 2008 to use tap wa-ter instead of groundwater, and if they failed to do so, they were to be either relocated or shutdown. Local residents had previously complained that the laundries caused water shortages because of their excessive use of groundwater.[144] In March 2009, the wells of five companies were shut off, with the government intending to close the wells of 48 companies in total.[145]

The government also plans to limit the amount of groundwater that can be extracted to 100 cu-bic metres a day.[146] Penalties for those who misuse groundwater will also be increased, with fines of up to Rp.1 billion (USD 109,000) and gaol terms of more than six years.[147] Previously many manag-ers of tall buildings exceeded their quota because the fine was less than the tariff for piped water.[148] However, by mid December 2008, 1,000 companies in Jakarta were facing over Rp.5 billion (USD 545,000) in fines for the misuse of groundwater. Some company representatives have complained about the fines, stating that they have to use groundwater because the piped water network fails to meet their needs.[149]

The Department of Energy and Mineral Resources has also asked the two water operators to increase their water supply. Palyja, which operates in the western side of Jakarta, has installed booster pumps to serve groundwater and tap water subscribers in the Mega Kuningan business district

---

[140] Fery Firmansyah, 'Tarif Air Tanah Jakarta Akan Dinaikkan' ('Groundwater Tariffs will be Raised'), *Tempointeraktif*, 27 February 2009, available at http://www.tempointeraktif.com/hg/ tata_kota/2009/02/27/brk,20090227-162358,id.html.

[141] Peraturan Gubernur No.4554/1999 (Regulation of the Governor No.4554/1999).

[142] Fery Firmansyah, note 141 above.

[143] Articles 9.2(b) and 12 of the Contracts.

[144] Anonymous, 'Laundries Behind Time to Install Tap Water, Waste Disposal Systems', *Jakarta Post*, 26 November 2008, available at http://www.thejakartapost.com/news/2008/ 11/26/laundries-behind-time-install-tap-water-waste-disposal-systems.html and Anonymous, 'Most Laundries Get Tap Water Installed', *Jakarta Post*, 27 November 2008, available at http://www.thejakartapost.com/news/2008/ 11/27/most-laundries-get-tap-water-installed.html.

[145] 'West Jakarta Closes Deep Wells to Conserve Groundwater', *Jakarta Post*, 17 March 2009, available at http:// www.thejakartapost.com/ news/2009/03/17/west-jakarta-closes-deep-wells-conserve-groundwater.html.

[146] *Berlebihan, Eksploitasi Air Tanah di Jakarta* (Excessive, Groundwater Exploitation in Jakarta) Suara Pembaruan, 18 November 2008.

[147] 'Govt Told to Monitor Groundwater Exploitation', *Jakarta Post*, 1 July 2008, available at http:// www.thejakartapost.com/ news/2008/07/01/govt-told-monitor-groundwater-exploitation.html.

[148] 'Pidana Bagi Penyedot Berlebihan' ('Criminal Sanctions for Excessive Extractors'), *Kompas*, 17 April 2008, available at http://asosiasi .org/tbk/2008/10/pidana-bagi-penyedot-berlebihan/.

[149] '1,000 Firms Fined for Overusing Groundwater', *Jakarta Post*, 12 December 2008, available at http:// www.thejakartapost.com/ news/2008/12/12/1000-firms-fined-overusing-groundwater.html.

in South Jakarta, increasing the water supply from Palyja from 30 litres per second to 50 litres per second.[150] Data indicates that there are fifteen groundwater users extracting 50,697 cubic meters per month (or 19.6 litres per second) of groundwater from 24 deep wells in the area. The groundwater users are all Palyja customers.[151]

CONCLUSION

Piped water supply in Jakarta is characterised by poor levels of access and quality. Reliability, limited coverage of the piped network, the low cost of groundwater, and water quality are important factors in determining consumer preference for groundwater.[152] This preference for groundwater has led to excessive groundwater use and theft, which is causing significant land subsidence, pollution and salinisation of aquifers and increased levels of flooding. The impact is so severe, the World Bank is predicting much of Jakarta will be inundated by seawater in 2025, rendering at least one third of the city uninhabitable and displacing millions of people.

In its effort to address this impending disaster, the government has taken the important step of revising its water management legislation so that it now adopts a holistic approach to water management, emphasises the need for efficient water use and recognises, although in a general sense, that water has a social, cultural, environmental and economic function and value. All these are positive developments. The government has also begun a public awareness campaign regarding the importance of groundwater.

However, reduction of groundwater use is generally judged to require an increased quantity and improved quality of piped water supply from surface water.[153] As discussed above, Jakarta's piped water system suffers numerous problems, including low service coverage, high water losses, low water quality and poor reliability. The government has embarked on an ambitious long term plan to improve the quality of raw water supplied to the piped water network, but is yet to develop long term policies for improvement of the network itself. The question therefore remains, has the government done enough and will it implement its commitment to sustainable groundwater use, or will groundwater use continue unabated making Jakarta the next lost city of Atlantis?

---

[150] The head of the groundwater and mineral management subagency Dian Wiwekowati, quoted in Adamrah, note 22 above. Aetra is planning to install a new booster pump in Cilincing, North Jakarta and to upgrade their booster pumps in Sungai Bambu and Sumur Batu in North and Central Jakarta in early 2009. *See* 'Water Quality Tops Complaints in Consumer Survey', *Jakarta Post*, 22 December 2008, available at http://www.thejakartapost.com/news/2008/12/22/water-quality-tops-complaints-consumer-survey.html.

[151] *See* Adamrah, note 22 above.

[152] Yusman Syaukat and G.C. Fox, 'Conjunctive Surface and Ground Water Management in the Jakarta Region, Indonesia', 40(1) *Journal of the American Water Resources Association* 241 (2004).

[153] *Id.* at 242.

Reuters Summit-Jakarta Sinks as Citizens Tap Groundwater

Sugita Katyal

JAKARTA, Oct 7 (Reuters)—It's one of the fastest-growing megacities in Asia. But some doomsters predict large parts of Indonesia's coastal capital could be under water by 2025.

The reason? Unchecked groundwater mining.

"Goundwater extraction is unparalleled for a city of this size," Almud Weitz, regional team leader of the World Bank's water and sanitation programme, said in an interview for Reuters Environment Summit.

"It's like Swiss cheese. People are digging deeper and deeper and so the city is slowly, slowly sinking. That is why tidal floods are occurring in poor areas on the coast."

Jakarta is one of Asia's more densely populated cities, but experts say it has one of the least developed piped water networks, pushing many residents as well as mushrooming megamalls and skyscrapers to increasingly suck out groundwater.

According to some estimates, Jakarta has a water deficit of about 36 million cubic metres (1.28 billion cubic feet) a year and much of the groundwater is contaminated with faecal matter because of leaky septic tanks.

As the city of around 10 million sinks and sea levels rise because of climate change, Jakarta has become more vulnerable to flooding and the threat of severe tidal surges remains grave.

In recent years, Jakarta, a city criss-crossed by 13 rivers and many canals built by its former Dutch rulers, has been devastated by massive flooding triggered by tropical rains and the incursion of sea water.

A study by a Dutch consultant for the World Bank showed that by 2025, the city could be between 40 and 60 centimetres lower than it is now, if nothing is done to check the crisis.

"An ever-growing population, densely populated residential areas, rapid infrastructural development, a diminishing number of green areas and catchments, plus six months of near-constant rain—you have a recipe for flood disasters which literally paralyse the city," the World Bank said in a statement when the study was released in April.

The Bank is supporting a flood management initiative with the local government.

Swathes of the teeming city were swamped and Jakarta's main airport was shut for hours earlier this year following heavy flooding caused by the combination of unusually high tides and the effects of subsidence from excessive extraction of groundwater.

QUEEN OF THE EAST

All this is in sharp contrast to the time when Jakarta, once known as the "Queen of the East", was renowned for its picturesque colonial houses, tropical tree-lined streets and canal network.

Today, the city is dotted with skyscrapers towering over run-down buildings and slums and is saddled with a host of problems, such as chaotic traffic, choking pollution and a massive influx of jobseekers each year.

Armi Susandi, a meteorologist at the Bandung Institute of Technology, who has researched the impact of climate change on Indonesia, estimates the submersion rate in the capital would be 0.87 cm a year, which is higher than an estimated average sea level rise rate of 0.5 cm a year until 2080.

Experts say the depletion of ground water has also allowed sea water from the Java Sea to seep into coastal aquifers, making the already filthy water saline.

Jakarta resident Maria Achmad said she and her family had been using ground water since 1995 because the piped water supply was erratic.

She was aware Jakarta was slowly sinking because of extraction of groundwater, but said her family didn't have a choice.

"I am concerned and worried about it. The Jakarta government should take action to limit the use of groundwater, especially by big restaurants and hotels which are the majority of groundwater users," she told Reuters."

"Second, we hope the government can fix the piped water system."

CLEARING THE CANALS

Faced with severe annual flooding, the Jakarta government has launched a multi-pronged plan that includes dredging the existing network of garbage-clogged canals and building a massive $560-million canal to stop the city from being swamped each year.

"We hope in two or three years we will manage to dredge all canals, revitalise the canals and, combined with the new canal, bring down flooding," Jakarta's governor, Fauzi Bowo, told Reuters.

"In previous years, when supply was short, a lot people tended to pump as much as possible. But now we have limited the usage of deep ground water because this is the main reason for land subsidence."

The river-dredging project overall would help reduce the flooded areas in Jakarta by up to 70 percent, but would still leave some parts of North Jakarta prone to flooding, Risyana Sukarma, a senior infrastructure expert with World Bank Indonesia, said on a Bank website.

"It is hoped that the dredging measures could return Jakarta floods to the previous cycle of once every 25 years provided that regular dredging maintenance takes place, and actions are to be taken to better manage solidwaste collection," Risyana said.

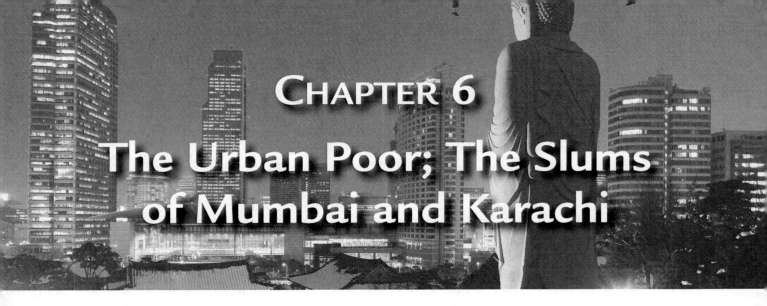

CHAPTER 6
The Urban Poor; The Slums of Mumbai and Karachi

Mumbai: The Redevelopment of Dharavi

Aaron Windle

Following an impressive accumulation of awards for the hit film "Slumdog Millionaire", the world's attention has been turned to the sprawling slums of India which comprised a large portion of the film's setting. The film depicts the squalid conditions and struggle for survival in the Dharavi slum, but ultimately leaves the audience with a sense of optimism from its people and desire to believe things may eventually turn out for the better there.

While the ending is certainly heart-warming, it is more to the benefit of the globally neglected Dharavi residents to take a critical perspective and delve into the Indian government's plans for land use, infrastructural development and proposed future housing schemes for the "hutment" residents. But in an effort to realize lasting positive change and provoke progress, the indigenous economy must not be ignored. The hectic pace of economic activity in the urban slums of India gives jobs and livelihood to the millions that gravitate towards it from all parts of India. A highly productive and thriving economic hub, the Dharavi slum is an interesting study in ingenuity, perseverance, and hope and its future is an interesting study in the politics of land use.

MUMBAI

Mumbai's development is relatively recent having sprung up from marshy lands occupied by a thriving fishing industry to a city of 16 million people in just 300 years. The earliest fishing villages and mangrove marshes were joined by European traders and, in later centuries, the swelling fortunes of the British in India and the growing importance of Mumbai as a port and trading center drew waves of immigrants looking for a better life.

Today Mumbai is the commercial and entertainment center of India, generating 5% of India's GDP and accounting for over 25% of industrial output, around 40% of maritime trade, and an astounding 70% of capital transactions to India's economy. However, juxtaposed against a backdrop of towering business buildings and economic renovation is Dharavi, one of the world's largest slums and a difficult situation to ameliorate from anyone's perspective.

THE SLUM

Dharavi was once an isolated settlement on the outskirts of Mumbai. It was unwanted land, rendered useless (due to the difficulty in traversing its swamps and marshes), except for its wildlife and the fishing village located near it. Since then, Dharavi has developed into a crowded collection of over 80 neighborhoods. These neighborhoods house an approximate 500,000 to 1,000,000 residents; however no recent (or reliable) population statistics are available. A 1986 survey by the National Slum Dwellers Federation (NSDF) counted 530,225 people (106,045 households) living in 80,518 structures. These numbers have almost certainly increased as Mumbai currently takes in an estimated 230 newcomers a day. Dharavi, as one of the biggest economic centers, is likely to have absorbed a sizeable proportion of that incoming migration.

It is difficult to grasp the density of Dharavi. While the total area is an estimated 550 acres, just short of one square mile, a recent survey by the Kamla Raheja Vidyanidhi Institute of Architecture (KRVIA) established that one central area of Dharavi, the Chamra Bazaar, contained densities of up to 18,000 people per acre. Assuming a population of approximately 700,000, the population density in Dharavi is 11 times as dense as Mumbai as a whole and more than 6 times as dense as daytime Manhattan without all the niceties of consistently running water, electricity, sewage solutions, roads, or healthcare. There is one municipal hospital and one toilet for every several hundred people, causing rampant disease and a paltry standard of living for residents.

ONE MAN'S TRASH . . .

Dharavi literally overlooks the Bandra-Kurla Complex which is the new financial and commercial center in Mumbai. In a city bound by water and already fighting to accommodate millions of inhabitants, the centrally located Dharavi represents substantial economic value and there is much pressure from developers and builders to free the land for commercial development that has been decades in the making.

Back in 1971, the Indian government enacted the Maharashtra Slum Areas (Improvement, Clearance, and Redevelopment) Act which drew an operational framework for preparing and submitting proposals for the modification of the development plan of Greater Mumbai and also created the Slum Rehabilitation Authority (SRA). According to the act the SRA was "… declared as a planning authority, to function as a local authority for the area under its jurisdiction. [The] SRA has been empowered to prepare and submit proposals for modification to the Development Plan of Greater Mumbai" according to their website.

In 2004, the Maharashtra state government accepted a $3-billion proposal submitted by Mr. Mukesh Mehta for the redevelopment of Dharavi. In a 2008 LA Times article he described the area by saying, "You're talking of a location that's fantastic. This is the only location in Mumbai where I can bulldoze 500 acres of land and redesign." Despite the fact that his redevelopment plan was adopted, it has been subject to perpetual debate and delay due to differences of opinion from residents. Some slum residents believe the redevelopment will provide a chance for them to live in dignity in proper apartments while others feel it is just another in a long line of disingenuous to oust them from their homes and lifestyle.

Mehta further stated that his goal is to "create a brand-new beautiful suburb complete with green space, schools, hospitals and reliable public services such as sanitation," things which Dharavi residents sorely need. Many locals feel the plan is ambitious for a place like Dharavi where failed examples of redevelopment can be readily found in the form of buildings abandoned halfway through construction and others in disrepair because of mismanagement and corruption.

THE SRA PLAN

According to the SRA website, several phases of redevelopment will transpire. Among other minor details, "the slum dweller whose name appears in the voter's list as of January 1, 2000 and who is the actual occupant of the hutment is eligible for rehabilitation. Each family will be allotted a self contained house of 225 ft² carpeted area free of cost. The eligible slum dwellers appearing in Annexure II certified by the Competent Authority will be included in the rehabilitation scheme. Eligible slum dwellers will be given rehabilitation tenements in Dharavi.

"During the implementation of this project, Dharavi residents will be provided with transit tenements, in close proximity of Dharavi or in Dharavi itself. The developer will bear the cost on account of rent of the transit tenements, but the cost of expenditure of consumables like water, electricity, telephone etc. will have to be borne by the slum dwellers."

The plan also includes improvements to infrastructure such as wider roads, electricity, ample water supply, playgrounds, schools, colleges, medical centers, socio-cultural centers, etc. For proper implementation, Dharavi has been divided into 10 sectors and sectors will be developed by different developers. The total duration of this project is expected to be between 5 to 7 years.

AND THE EXISTING LOCAL ECONOMY?

Fortunately the planning authorities are taking the local economy into consideration when planning for the future employment of the slum dwellers. One estimate places the annual value of goods produced in Dharavi at $500 million. Commercial and home manufacturing businesses offer employment for a large share of its population as well as for some workers living outside the slum. For the various industrial units in Dharavi, it is being proposed that non-polluting industrial businesses be retained. Current slum cottage industries include large-scale recycling, leather tanneries, metal work, woodwork, machinery manufacturing, printing, garment finishing factories, and other small scale factories producing a wide array of manufactured goods mostly for local consumption but for export as well. There is also local manufacturing of shoes, luggage, and jewelry. Almost any economic activity that can be accommodated within the confines of small, corrugated aluminum huts exists in Dharavi.

In addition, the SRA states that "All the currently established businesses and manufacturing units will be incorporated into the plan and will be provided with modern technical and economical strategies for sustainable development." How exactly these cottage industries will be maintained in a new, ostensibly less polluted housing arrangement has yet to be delineated, but at least the SRA and developers have not completely ignored the long-term economic plight of the lower caste in Dharavi by making some allowances for current revenue streams.

FROM HERE ON OUT

It is certainly uncommon in the world of real estate to contrast slum land with an estimated value of $10 billion to the poverty of its residents, many of whom live on close to $1 per day. In total, the proposed development projects throughout the Dharavi slum, hopefully to be undertaken by local developers and government housing entities in short fashion, amount to $3.4 billion. While no one would envy the lives of the slum's residents kept afloat on an economy of $500 million inside Dharavi's makeshift city limits, relocation outside of the slum might not improve the residents' economic situations. Most 21st century land use notions are poorly equipped to create communities that, on a massive scale, both house the poor and accommodate their self-sufficient economies. Redevelopment certainly has its deficiencies; in this instance, though, it may also have its casualties.

CHAPTER 7
Economic Growth; Tokyo, Japan—One of Three World Centers

Building World City Tokyo: Globalization and Conflict Over Urban Space

André Sorensen

Abstract. Japanese policy makers have, since their contact with the colonial powers in the mid 19th century, been acutely aware of the pressures and challenges of national survival in a globalizing world. In this sense, the Japanese experience of modernity has been deeply intertwined with, and is in important ways inseparable from the ongoing processes of globalization during the last century and a half. While their main response was to foster the growth of Japanese industrial, military and diplomatic power, one consistent theme has been the development of the capital city Tokyo as emblem of Japan as a civilized nation, location of national institutions, and center of economic power. This project, however, has long been an arena of considerable conflict between city builders and the residents of central Tokyo. The most recent conflict over the control of urban space in Japan's premier world city emerged in the last few years when major developers lobbied successfully for massive increases in allowable building volumes and heights in special regeneration areas, arguing that without further deregulation Tokyo would lose its competitive position in relation to Shanghai, Hong Kong and Singapore. This paper argues that in Japan an important feature of globalization and international competitive pressures has been their use by urban actors in disputes over the control of urban space, and examines this use of globalization debates in the competition between economic space and life space in Tokyo.

INTRODUCTION

"Competition between cities is international—to revitalize Tokyo is to revitalize Japan . . . if Japan is to prosper in an age of globalization, its cities must attract residents, goods, finance and information from around the world."

Mori Building Website, July 2002

One of the key contributions of urban studies to the globalization debate has been the concept of 'world cities'. The research agenda about world cities as first proposed by Friedmann and Wolf (1982), Friedmann's more developed hypothesis (1986) and the monograph by Sassen (1991) which proposed a triumvirate of leading 'global cities' (London, New York, and Tokyo) have informed a vast literature on globalization and world cities. The core idea was that a number of global processes including increased integration of world commodities, finished goods and financial markets, and growing interconnection through communications networks were likely to result in a convergence of economic structures, and that this would in turn have similar spatial and social impacts in diverse world cities. Looking at the leading 'global cities,' Sassen (1991: 4) found that indeed such parallel changes in economic base, spatial organization and social structure were occurring, and strongly asserted the primary role of global processes in those changes.

Subsequent research put more emphasis on the role of national and local actors and institutions in mediating, contesting, and shaping the particular products of economic changes in particular cities (Goetz and Clarke 1993; Knox and Taylor 1995; Eade 1997). The prevailing interpretation today is that while global forces are generally the more powerful, and affect all major cities depending on their degree of interconnectedness with the world economy, local histories and institutions do influence outcomes in particular places, which are diverse (e.g., Pacione 2001: 8).

Tokyo has played a lead role in globalization debates from the beginning. Sassen (1991) stressed the importance of Tokyo as a key node in the world economic system alongside London and New York, and a large body of work discerned a range of globalization induced changes to the Tokyo space-economy (Rimmer 1986; Douglass 1988; Fujita 1991; Douglass 1993). In particular, Tokyo's emergence as a key global command and control center, the increasing importance of financial, insurance and real estate industries in its economy and their consequent demand for central city office space were seen as having major impacts on the space economy of central city areas (Sassen 1991). Fujita (1991) argues that Tokyo's development into a world city is primarily a result of its flexible manufacturing prowess, which captured a major share of world markets, allowed the creation of vast trade surpluses, and encouraged the growth of financial services industries. That in turn increased the capital's dominance over the Japanese economy as well as intensified competition for space in central areas, exacerbating housing problems and other social tensions. Others described the adverse social impacts of urban restructuring including upward spiraling land prices and rents resulting from buoyant markets for office space, the displacement of population from central city areas, the uprooting of long-standing communities, and lengthening commutes (Douglass 1993: 88). He suggests that while economic growth had certainly brought Japanese people higher incomes and greater consumption, urban quality of life had not improved, and may even have worsened (Douglass 1993: 88). Machimura (1992, 1998) stresses the political conflicts arising from urban restructuring, and argues that the concept of world city has been used symbolically as a political level in domestic political conflicts over urban space.

Recent work has contributed case studies of specific globalization-related changes, such as the development of the waterfront subcenter on reclaimed land in Tokyo Bay, which was developed explicitly to advance Tokyo's world city ambitions (Saito 2003). Saito examines the political dynamics behind the subcenter project, and finds that the main player was the Tokyo Metropolitan Government, which saw its role as fostering Tokyo's emergence as a premier world city by creating an advantageous location for high tech office towers and high amenity living space. Saito's fascinating play-by-play description of the evolving political dynamics behind the project provides a useful glimpse into Japanese policymaking. In contrast to much of the Anglo-American literature on globalization and world cities, which sees market forces as the overwhelmingly dominant force behind spatial restructuring in world cities, in Saito's examination of the waterfront subcenter it was the State that dominated the process, not private enterprise. Following Hill and Kim (2000) he explains this strong role of the state as a product of Japan's history of "developmental state" policies.

In the present context his stress on the importance of the developmental state is significant. While there has been an extended debate among students of Japanese political economy of the relative importance of the state bureaucracy, political parties, and big business (see, e.g., Johnson 1982; Deyo 1987; Gao 1997), no one would argue that Japan is a liberal market economy in the pattern of the United States. Here it is not necessary to take sides in the debate over the precise degree of credit for Japanese economic success to assign the Japanese bureaucracy. It is important, however, to understand that Japan had a much different experience of modernization, industrialization and urbanization than any of the other developed countries, and that this contributed to the development of a highly distinctive planning culture. The Japanese leadership always saw the building of a strong State as the primary national project—a project to which the people should subordinate themselves—and which was commonly cast in a framework of fierce international competition in which Japan played the role of outsider catching up. While there were dramatic changes over the course of the 20th century, the focus of resources on developmental goals rather than public welfare has proven highly enduring, as shown below.

Far from the competitive pressures of globalization being a phenomenon new to the 1980s, therefore, in Japan they have been a driving force behind national policymaking for at least a century. In this competitive world environment, the Japanese leadership consistently put national development and international stature ahead of private welfare or even collective quality of life. This makes it somewhat difficult to distinguish a particular period during which globalization pressures begin to impact Japanese public policy. While it is often suggested that deregulation and the hollowing out of the welfare state are prominent manifestations of globalization, in Japan the welfare state was always thin, the state sector small, and as shown below, regulation of urban development weak. The priority was economic development, and urban planning was carried out primarily to foster that growth, not to help create or maintain amenities or improve the quality of life in urban areas (see Sorensen 2002). In Friedmann's (1988) terms, the priority was always "economic space" not "life space."

Friedmann's concept of the conflict between "life space" and "economic space" is particularly helpful for understanding Tokyo's restructuring conflicts. Tokyo was already a giant city of over a million during the feudal era two centuries ago, and the current structure of land uses is still strongly influenced by its long urban history (see Cybriwsky 1998; Sorensen 2002). Urban restructuring and redevelopment in the early postwar period was carried out primarily along major new arterial roads and at nodes in the public transport system. This bypassed many older traditional neighborhoods that are characterized by extremely fragmented land ownership patterns and narrow roads inherited from earlier periods. Apart from the real difficulty of land assembly for redevelopment in such areas, a building code regulation first passed in 1919 ties allowable building height to a slant plane drawn from the opposite side of the road on which a property fronts (Sorensen 2002: 116). Throughout Tokyo, but particularly in areas of narrow roads and small lots, this regulation severely limits the height of buildings and development capacity (Onishi 1994). It also helped to protect from redevelopment many inner city districts, which commonly maintained many of the characteristic features of traditional Japanese cities; narrow roads, small neighborhood commercial centers (shotengai) of family run shophouses with the shop at ground level and residence above, strong neighborhood organizations which managed the local shrine and festival, recycling collection, park maintenance and fire patrol among other functions (Bestor 1989). The central area of Tokyo thus retained many close-knit residential communities, which were geographically distinct and separated from the main areas of modern highrise development. As shown below, however, such areas of "life space" have come under repeated pressure for profitable redevelopment to "economic space" since the 1960s.

While this has been a long-term struggle between the forces of economic change and community stability, however, it is only in the 1980s and 1990s that the language of globalization and world cities began to be used to legitimize and explain these conflicts, as shown below. Section 2 outlines the weak planning system of the rapid growth period in the 1950s and 1960s, and describes several key planning protections won by progressive political forces and social movements in the 1960s and 1970s. Section 3 outlines the swings from deregulation in the 1980s to re-regulation after the bubble to the most recent period of deregulation from the late 1990s. Section 4 summarizes the main findings.

SHIFTING THE BALANCE BETWEEN DEVELOPERS AND COMMUNITIES, 1960s–1970s

During the rapid economic growth period of the 1950s and 1960s Japan had an extremely weak system of urban land planning and development control. This assertion seems to contradict the widely held perception of a "Japan Inc." that was a highly efficient, technocratic, well-planned developmental state which effectively mobilized national resources in pursuit of economic growth (Johnson 1982). In fact, while the latter description is certainly an oversimplification of the case (for other interpretations see, e.g., Muramatsu and Krauss 1987; Calder 1988), it is widely agreed that during the rapid growth period an "iron triangle" of Liberal Democratic Party (LDP), central government bureaucracy, and big business worked effectively to mobilize available national resources, and produced spectacular economic growth. At the same time, this focus of the nation's resources on industrial development and capital formation resulted in ongoing shortages of investment in social overhead capital. As Honjo put it: "The conditions under which Japan developed were so severe that it was impossible to do more than the bare minimum. The accumulated capital was always mobilized for investment in productive sectors, and an urban development policy focused on infrastructure was promoted . . . The housing supply was left to the private sector, and only during emergencies such as natural disasters were public measures initiated or expanded" (Honjo 1984: 28). The role of city planning was seen as the supply of infrastructure for economic growth: highways, ports and airports, industrial water supply, serviced industrial sites, and low-cost public housing for the workers who migrated in their millions to the cities (Morimura 1994). Little money was spent on residential areas and low priority was put on investment in the more discretionary public goods such as parks, local roads or sidewalks. Instead of creating a system to regulate private urban development, the state provided a range of basic infrastructure such as water supply and arterial roads, while encouraging private investors to provide other needed urban investment such as in electrical generation, commuter railways, and housing development.

The weak planning system was thus a result not of a lack of state capacity, as in many developing countries today, but of a narrowly focused urban policy which prioritized economic development. The main planning tool was a weak zoning system, with only four zones, residential commercial, industrial and quasi-industrial. Housing could still be built within industrial and commercial zones, and small-scale industry within residential and commercial zones. Within zoned areas land development was as-of-right, with no requirements for basic urban infrastructure before land development, no subdivision control, nor any minimum housing standards. The result was extremely rapid growth of haphazard un-serviced sprawl combining housing, commercial and industrial uses at high densities. In primarily residential areas much housing was built along narrow, unpaved private roads, and large areas were built up without municipal sewers, parks, piped gas supply or even sidewalks (Sorensen 2001a).

This weak planning system was very much the product of strong central control. This is partly because local governments are highly dependent on grants from the central government, and that funding was tightly focused on the developmental priorities noted above. It is even more a result, however, of the fact that planning law was written and interpreted by central government ministries, and local governments had no legal authority to go beyond the parameters set by national legislation. For example, because the national government had set no minimum housing standards, minimum plot sizes, or minimum infrastructure requirements for land development, municipal ordinances that attempted to introduce such measures could not be legally enforced, and local governments frequently lost court challenges to them (Jain 1991). In this way central government effectively limited the planning tools available to local governments.

While in the early post-war period there was a broad agreement that economic growth was a top priority, the success of that project created other pressing problems that effectively shattered the consensus around growth during the 1960s. Large-scale development of heavy and chemical industries intermixed with or in close proximity to residential and commercial areas, combined with almost non-existent pollution controls, resulted in a severe environmental crisis. Pollution of air, water and food supplies was directly related to the spread of environmental diseases, and hundreds died, thousands suffered debilitating and painful diseases or were born deformed, and hundreds of thousands suffered from asthma and other chronic pollution-related ailments (see Huddle, et al. 1975; McKean 1981; Ui 1992; George 2001). Eventually, large numbers of local environmental protest movements developed to lobby for better pollution control regulations and against industrial development (McKean 1981; Upham 1987).

This conflict extended also to urban planning policy. Krauss and Simcock (1980: 196) suggest that there was an "explosion of protest in urban and suburban areas" against industrial plants and highway interchanges, and demanding that local governments provide essential services such as sewers, parks and sidewalks. As Samuels (1983: 190) described it, "The left came to power by convincing enough of the electorate that the conservative central government and their allies in the localities were responsible for the pollution, the lack of social programs, and the support of business interests at the expense of residents." The growing political strength of progressives in both local and central government elections scared the ruling LDP into passing new city planning legislation in 1968 (Calder 1988: 405). The New City Planning Law of 1968 was the first major post-war swing of the pendulum towards greater emphasis of urban planning on urban quality of life and tighter regulation of development, and it generated high hopes that local governments would finally have the tools to be able to control land development and improve urban environments.

Although the new planning system encountered serious difficulties in managing urban growth on the fringe (see Nakai 1988; Hebbert 1994; Sorensen 1999; Sorensen 2001a), the new zoning system did have a significant impact on development in existing built-up areas. The key factor was the introduction of the Exclusive Residential Zone #1, which was the first land use zone meaningfully restricting land uses. In the new zone not only was land use restricted to residential uses and related compatible land uses (e.g., churches, small-scale retail), but a strictly enforced absolute height limit of 10 meters was also imposed. The height limit was important. The Japanese building code had since the prewar period maintained a strictly enforced absolute height limit of 30 meters on all buildings because of concerns about earthquakes. However, the limit was abolished by the 1970 revision to the Building Standards Law because of dramatic improvements in engineering technology using steel reinforced concrete. As a result, extensive low- and medium-rise areas were suddenly ripe for redevelopment into higher-rise buildings, and the 1970s saw a rush to inner-city condominium building, the so-called "manshon boom."

The condominium boom created severe conflicts in many areas, in part because the high-rise buildings brought with them increased local congestion and noise, but even more because they almost always blocked direct sunlight to neighboring houses for part of the day. In Japan direct sunlight is an essential aspect of residential quality of life, houses have long been routinely oriented towards the sun, and an important part of housewives' daily routine—even today—is to hang bedding out in the sun to air it out. Where buildings block the sun, residential amenity is permanently impaired. In residential areas throughout Tokyo local citizens' groups organized to oppose the building of high-rise apartment buildings which blocked the sunlight from surrounding houses (Ishizuka and Ishida 1988: 30).

As the only land use zone that retained any height control function was the Exclusive Residential Zone #1, designation of such zones became a highly contested process, as it effectively barred redevelopment to high rises. In the early 1970s, the Tokyo Prefecture and most Tokyo wards were controlled by progressive administrations, and in many cases the rezoning to the new zoning system gave priority to the protection of existing traditional residential communities. In 1972, again responding to electoral pressure and increasingly vocal citizen movements, the government passed an amendment to the Building Standards Law that allowed the creation of Height Control Zones with a maximum building height of 10 meters that could be designated over other land use zones. Such height control zones were widely designated by sympathetic ward governments to protect residential areas throughout central Tokyo and became an increasing source of conflict later, as discussed below.

A third factor that served to protect low-rise residential areas from high-rise redevelopment was directly the result of the activities of the citizen movements of the early 1970s. A Tokyo-wide organization of sunshine rights groups won a series of court cases against developers of high-rise buildings, forcing them to compensate the neighbors they had deprived of light. In 1972, the Supreme Court determined that Article 25 of the constitution, which guarantees "minimum standards of wholesome and cultural living," protected the right to sunshine, and that infringements of sunshine rights were liable for damages (McKean 1981: 113). Then in 1973 the citizens' movement drafted and presented their own proposal for a sunlight protection ordinance *(Hiatari Jôrei)* to the Tokyo Metropolitan Government. As Ishida and Ishizuka suggest: "This was an epochal development in the citizens' movement in that it progressed from simply opposing things to actively proposing policies" (Ishizuka and Ishida 1988: 30). Through the 1970s a large number of sunshine rights cases were won against offending builders and even against the government in the case of an elevated expressway.

In this context the government worked quickly to revise the building regulations. The level of uncertainty about what was permissible and what would incur liability to pay damages was high enough that builders ran an unacceptably high risk of losing court battles with neighbors, even where they had complied with existing regulations. The Ministry of Construction (MoC) drafted a revision to the building standards law that included many of the proposals of the citizens' movement, and those revisions came into effect in 1976. The revision required all local governments to draft their own sunshine standards that specified the minimum hours of unimpeded sunlight cast to the north of new buildings on the winter solstice when the sun is at its lowest.

These three changes, the creation of Exclusive Residential #1 zones, the ad hoc height control zones, and the sunlight preservation regulations, significantly changed the balance of power between urban residents and the development industry, and permitted local governments to increase their controls over development in existing built up areas. For the first time residential neighborhoods had some leverage to protect themselves from redevelopment initiatives. That this new regime really did represent a significant shift in the power balance is suggested by the fact that from the early 1980s the central government, urged on by the development industry, started to work hard to weaken these regulations to permit more redevelopment in central Tokyo.

FROM DEREGULATION IN THE 1980s TO RE-REGULATION IN THE 1990s AND INCENTIVES IN 2000

In the early 1980s the urban policy climate shifted abruptly. The new Prime Minister Nakasone, inspired by the neoliberal policies of Thatcher and Reagan, promoted policies of deregulation, privatization and fiscal retrenchment. A key rationale for deregulation was to enhance Japan's international economic competitiveness, and urban planning deregulation efforts were increasingly cast in terms of the economic importance of World City Tokyo in a globalizing world. At the Tokyo Metropolitan Government level the conservative administration of governor Suzuki strongly supported a "world city" strategy for Tokyo's glorious future as an international city (Tokyo Metropolitan Government 1987), although Machimura (1998) argues that this use of the term was primarily symbolic, designed to impress and gain support for the deregulation program.

Deregulation of city planning regulations was carried out throughout the 1980s. While Nakasone's neoliberal agenda closely followed those of Thatcher and Reagan, urban policies were explicitly oriented towards the remodeling of Tokyo into a competitive World City. Hence, a central and early part of Nakasone's deregulation campaign focused on encouraging redevelopment in central Tokyo. For example, one of the first actions of the Nakasone government upon taking office in 1982 was to direct the MoC to review the zoning of all the areas in central Tokyo that were zoned Exclusive Residential #1. The idea was to rezone them to Exclusive Residential #2 in which high-rise buildings can be built, so that Tokyo could begin to look more like New York (Hebbert and Nakai 1988: 386; Miyao 1991: 132).

Then in March of 1983 the Ministry of Construction ordered all local governments to encourage development by relaxing regulations. Specifically, they were to increase the ratio of building volume to lot size, rezone residential zones to commercial, and weaken various restrictions on urban fringe land development (Hayakawa and Hirayama 1991: 153). The central government also strongly pressured local governments to weaken or abolish their non-statutory "Development Manuals" which specified required levels of contribution to public infrastructure to get a development permit.

The relatively low intensity of land use in central Tokyo compared to other developed country major cities was seen as a problem. For example, the Tsukuba National University economist Takahiro Miyao argued that planning restrictions limiting the height of buildings in much of central Tokyo were a key urban problem because they made it difficult to redevelop central city areas to more intensive uses such as high-rise condominiums. He argued for further deregulation, suggesting that the vigor of the private sector needed to be freed from excessive planning regulations in order to "take full advantage of the vitality in the metropolitan regions" (Miyao 1987: 58–9). Specifically, deregulation should ensure that "excessive restrictions of the residential area development guidelines by local municipalities would be corrected" (Miyao 1987: 59).

Even though most ward governments successfully resisted pressure to abolish the Exclusive Residential #1 Zones in central Tokyo, the list of national planning deregulations was long, and these had powerful impacts on urban redevelopment patterns (Hayakawa and Hirayama 1991; Otake 1993; Inamoto 1998). One particularly important new planning measure for central area redevelopment was an incentive to provide more inner city public space. Modeled after New York's Plaza Bonus system, it rewarded developers with extra height and floor space allowances in return for the provision of public open space or plazas at ground level. This was a logical measure in a city as crowded as Tokyo, with a shortage of public open space, particularly in core areas. Developers were eager to use it. Even though there were no absolute height restrictions in much of central Tokyo, floor area ratios, slant plane restrictions and sunlight ordinances still put serious limits on building heights and volumes. The plaza bonus measure allowed them to negotiate significant increases in height and bulk over and above those limits.

At the Tokyo Metropolitan Government level the conservative administration of governor Suzuki actively supported its own vision of polycentric development within the greater Tokyo area (Sorensen 2001b). Of the various subcenters, the most closely related to the world city project was the waterfront subcenter to be built on reclaimed land in Tokyo Bay (Saito 2001). Promoted initially as Tokyo Teleport Town, which was planned to function as Japan's main international communications gateway, the concept was to provide a major new supply of high quality office space in a high tech cluster. Other major infrastructure projects pursued by the Suzuki administration such as new subway lines, water supply projects, wastewater treatment plants, and garbage incinerators were also designed to facilitate central area redevelopment and intensification.

Planning deregulation, fiscal stimuli, and infrastructure spending were successful beyond expectations. In the second half of the 1980s the Japanese economy boomed and redevelopment of central Tokyo proceeded apace. While at the time the boom seemed very impressive, and "triumphalist" predictions abounded that Japanese economic output would soon surpass that of the US, in retrospect the boom was an unmitigated disaster. Land prices soared, first in central Tokyo and then in the rest of the country, resulting in sharp increases in inequality as those with land assets gained and those without found themselves ever farther away from owning their own home (Tachibanaki 1992; Noguchi and Poterba 1994). The frenzy of real estate investment amid continuously rising land prices created powerful incentives for the high-rise redevelopment of existing low-rise inner city residential areas.

While there is no doubt that in many cases small inner city landowners were happy to sell their property at very high prices and move out to the suburbs, in many other cases owners wished to stay in communities where they had long ties. Even more reluctant to move were tenants who often had low rents and legal protection against eviction. In cases where owners were reluctant to sell, or tenants preferred not to move, land assembly became the work of "land sharks" (jiageya), or real estate gangsters who used intimidation, threats and violence to encourage people to leave (Hayakawa and Hirayama 1991: 156; Cybriwsky 1993: 140). Another tactic that was particularly harmful to inner city communities was the buying up and closure of privately operated public baths that functioned both as an essential public facility (in areas where many houses had no baths of their own) and as a key neighborhood meeting place. The number of public baths declined from 22,650 in 1964 to 39 by the end of 1986 (Douglass 1993: 98). There is no doubt that the real estate investment explosion of the 1980s greatly heightened conflict over space in central Tokyo, and that deregulation weakened the ability of communities to maintain life space against the encroachment of economic space. In order to increase profits to real estate developers, inner city communities were destroyed. Housing problems increased sharply, particularly for the very poor (Kodama 1990; Watanabe 1992; Oizumi 1994). Unfortunately, the boom also seriously damaged the Japanese economy. When the bubble burst, the nation's financial system almost collapsed under the weight of bad real estate loans, and the country has suffered economic stagnation since 1991 up to the present.

By the late 1980s there was increasing public pressure to address the land inflation crisis. The ongoing process of deregulation and government down-sizing was, at least temporarily, derailed, and the committee in charge of planning and land deregulation in particular was forced to do an about face and start promoting strengthened land development and speculation controls as detailed by Otake (1993). In the late 1980s and early 1990s there were several moves in the direction of re-regulation of land development, especially the Basic Land Law of 1989. The Law declared that: "First, public welfare should be given priority over private profit in the ownership and use of land. Second, land should be used in a proper and orderly fashion. Third, land should not be an object of speculation. Fourth, landowners should return a part of their profits to the public through imposition" (Oizumi 1994: 210).

Another step was a significant strengthening of the City Planning Law, including provisions for Master Planning and improved zoning regulations among others (Watanabe 1992; Oizumi 1994: 211; Sorensen 2002: 302). While there has been considerable debate about the effectiveness of many of these measures, it is clear that the excess of the bubble economy period had effectively undermined the political viability of the deregulation argument. As a result, the pendulum swung again in the early 1990s in favor of greater public and governmental support for stronger planning and the tighter regulation of land development.

More recently, the extended recession of the 1990s has again tipped the political advantage in favor of the property development industry. In particular, the financial system has been on the brink of collapse because of massive bad loans resulting from real estate deals during the bubble, the crash of equities values, and the continuing decline in Japanese land prices since 1991. Hence, government policy has given priority to stabilizing the financial system. This has included major bailouts to the banks and huge spending on infrastructure projects to inject funds into the construction and development industries.

Perhaps most important in the present context have been moves to deregulate the property development industry to make land development profitable once again to compensate for bubble-related losses. During the 1990s a series of changes by the central government to the Building Standards Law have allowed significant increases in allowable building heights and volumes in order to make urban redevelopment more profitable. The most recent change results from the passage in April of 2002 of the Special Urban Regeneration Act by the Koizumi government. This act established an urban regeneration office within the national cabinet. This office has the authority to designate Urban Regeneration Areas in which greatly weakened development regulations will apply and FAR bonus systems such as for Plazas can be permitted directly by central government instead of requiring local government consultation. This avoids the head-on confrontation with local governments over rezoning that proved so problematic in the 1980s. The main advocates of these measures have been the largest property development companies that stand to profit considerably from floor area bonuses and deregulation. These companies have argued strongly that competition from Singapore, Shanghai and Hong Kong requires urgent measures if Tokyo is to remain competitive in the global property market. The main opponents have been local governments and local communities that have seen a significant erosion of local legal powers to regulate inner city redevelopment.

By giving cabinet authority to designate regeneration areas, the central government has made it much easier and faster to grant approval for huge FAR bonuses for redevelopment of inner city sites into high-rise global space. This sidesteps the protests of local residents and the sometimes extended processes of public consultation demanded by local governments. The government hopes this will contribute to the high-rise redevelopment of the low-rise areas of inner Tokyo, provide more housing and office space to allow Tokyo to compete better with other world cities, and restore land development companies to profitability.

CONCLUSIONS

There are different ways of interpreting this history. It could be argued that Tokyo provides a classic case of globalization impacts on urban change. The last two decades have seen the unremitting encroachment of life space by economic space, repeatedly sponsored and encouraged by the central government in the name of building a competitive global city. While Japan's developmental state history makes it distinctive in many aspects of its relationships between State, society and market, it is possible to argue that in Japan one important impact of global economic integration and increased international competition has been increased pressure for the redevelopment of traditional residential areas to high-rise office space. A significant consequence has been the weakening of the planning protections inner-city communities had gained during the struggles of the 1970s.

On the other hand, it can also be argued that the concept of globalization in Japanese political discourse is primarily used a strategic tool to gain political advantage. In this sense, the stress on the need to compete with Shanghai and Hong Kong in the lobbying by property developers for weaker development regulations, more ad hoc FAR bonuses and greater public subsidy of infrastructure improvements can be interpreted as mere opportunism. The fact that the development industry has gained significant benefits from their use of the globalization threats does, however, indicate the continuing political usefulness of globalization/world city discourses. In the Japanese case, therefore, globalization does not seem to influence urban change primarily through foreign direct investment in urban (re)development, but by providing a convincing political argument for the weakening of local planning controls and participation processes.

What is clear is that since the high point of citizen mobilization over environmental issues and progressive control of local governments in the mid 1970s, communities in inner city areas have, on the whole, been fighting a losing battle against a resurgent property development industry that seeks to redevelop inner city neighborhoods into high-rise towers. Issues of quality of life and urban livability continue to be given short shrift, communities in inner city Tokyo continue to feel pressure, and life space continues to be redeveloped as economic space.

REFERENCES

Bestor TC (1989) Neighborhood Tokyo. Stanford University Press, Stanford Calder KE (1988) Crisis and compensation: public policy and political stability in Japan, 1949–1986. Princeton University Press, Princeton.

Cybriwsky R (1993) Tokyo. Cities 10(1):2–11.

Cybriwsky R (1998) Tokyo: The changing profile of an urban giant. Wiley, Chichester Deyo FC (ed) (1987) The political economy of the new asian industrialism. Cornell University Press, Ithaca.

Douglass M (1988) The transnationalization of urbanization in Japan. International Journal of Urban and Regional Research 12(3):425–454.

Douglass M (1993) The 'New' Tokyo story: Restructuring space and the struggle for place in a world city. In: Fujita K, Hill RC (eds) Japanese cities in the world economy. Temple University Press, Philadelphia, pp 83–119.

Eade J (1997) Living in the global city: Globalization as local process. Routledge, London.

Friedmann J (1986) The world city hypothesis. Development and Change 17:69–84.

Friedmann J (1988) Life space and economic space: Contradictions in regional planning, in life space and economic space: Essays in third world planning. Transaction Books, New Brunswick, N.J, pp 93–108.

Friedmann J, Wolff G (1982) World city formation: An agenda for research and action. International Journal of Urban and Regional Research 6(3):309–344.

Fujita K (1991) A world city and flexible specialization: Restructuring of the Tokyo metropolis. International Journal of Urban and Regional Research 15(1):269–284.

Gao B (1997) Economic ideology and Japanese industrial policy: Developmentalism from 1931 to 1965. Cambridge University Press, New York.

George TS (2001) Minamata: Pollution and the struggle for democracy. Harvard University Press, Cambridge.

Goetz EG, Clarke SE (eds) (1993) The new localism: Comparative urban politics in a global era. Sage, Newbury Park.

Hayakawa K, Hirayama Y (1991) The impact of the minkatsu policy on Japanese housing and land use. Environment and Planning D: Society and Space 9:151–164.

Hebbert M (1994) Sen-biki amidst Desakota: Urban sprawl and urban planning in Japan. In: Shapira P, Masser I, Edgington DW (eds) Planning for cities and regions in Japan. Liverpool University Press, Liverpool, pp 70–91.

Hebbert M, Nakai N (1988) Deregulation of Japanese planning. Town Planning Review 59(4):383–395.

Hill RC, Kim JW (2000) Global cities and developmental states: New York, Tokyo and Seoul. Urban Studies 37(12):2167–2195.

Honjo M (1984) Key issues of urban development and land management policies in Asian developing countries. In: Honjo M, Inoue T (eds) Urban development policies and land management: Japan and Asia. City of Nagoya, Nagoya, pp 15–35.

Huddle N, Reich M, Stiskin N (1975) Island of dreams. Autumn Press, New York.

Inamoto Y (1998) The problem of land use and land prices. In: Banno J (ed) The political economy of Japanese society, vol. 2, Internationalization and domestic issues. Oxford University Press, Oxford, pp 229–264.

Ishizuka H, Ishida Y (1988) Tokyo, the metropolis of Japan and its urban development. In: Ishizuka H, Ishida Y (eds) Tokyo: Urban growth and planning 1868–1988. Center for Urban Studies 3–35, Tokyo.

Jain PC (1991) Green politics and citizens power in Japan. The Zushi Movement. Asian Survey 31(5):559–575.

Johnson C (1982) MITI and the Japanese miracle, the growth of industrial policy, 1925–1975. Stanford University Press, Stanford.

Knox P, Taylor PJ (eds) (1995) World cities in a world system. Cambridge University Press, New York.

Kodama T (1990) The new aspects of housing problems in Tokyo. Osaka City University Economic Review 25(1):1–12.

Machimura T (1992) The urban restructuring process in Tokyo in the 1980s: Transforming Tokyo into a world city. International Journal of Urban and Regional Research 16:114–128.

Machimura T (1998) Symbolic use of globalization in urban politics in Tokyo. International Journal of Urban and Regional Research 22(2):183–194.

McKean M (1981) Environmental protest and citizen politics in Japan. University of California Press, Berkeley.

Miyao T (1987) Japan's urban policy. Japanese Economic Studies 15(4):52–66.

Miyao T (1991) Japan's urban economy and land policy. Annals of the American Academy of Political and Social Science (513 January):130–138.

Morimura M (1994) Change in the Japanese urban planning priorities and the response of urban planners 1960–90. In: University of Tokyo Dept. of Urban Engineering (ed) Contemporary studies in urban environmental management in Japan. Kajima Institute Publishing 8–24, Tokyo.

Muramatsu M, Krauss E (1987) The conservative policy line and the development of patterned pluralism. In: Yamamura K, Yasukichi Y (eds) The political economy of Japan, vol. I. The domestic transformation. Stanford University Press, Stanford, pp 516–554.

Nakai N (1988) Urbanization promotion and control in metropolitan Japan. Planning Perspectives 3:197–216.

Noguchi Y, Poterba JM (eds) (1994) Housing markets in the United States and Japan. University of Chicago Press, Chicago and London.

Oizumi E (1994) Property finance in Japan: expansion and collapse of the bubble economy. Environment and Planning A 26(2):199–213.

Onishi T (1994) A capacity approach for sustainable urban development: An empirical study. Regional Studies 28(1):39–51.

Otake H (1993) The rise and retreat of a neoliberal reform: Controversies over land use policy. In: Allinson G, Sone Y (eds) Political dynamics in contemporary Japan. Cornell University Press, Ithaca, pp 242–263.

Pacione M (2001) Urban geography: A global perspective. Routledge, London.

Rimmer P (1986) Japan's world cities: Tokyo, Osaka, Nagoya or Tokaido Megalopolis. Development and Change 17:121–158.

Saito A (2003) World city formation in capitalist developmental state: Tokyo and the waterfront sub-centre project. Urban Studies 40(2):283–308.

Samuels RJ (1983) The politics of regional policy in Japan: Localities incorporated? Princeton University Press, Princeton, New Jersey.

Sassen S (1991) The global city: New York, London, Tokyo. Princeton University Press, Princeton, New Jersey.

Sorensen A (1999) Land readjustment, urban planning and urban sprawl in the Tokyo metropolitan area. Urban Studies 36(13):2333–2360.

Sorensen A (2001a) Building suburbs in Japan: Continuous unplanned change on the urban fringe. Town Planning Review 72(3):247–273.

Sorensen A (2001b) Subcentres and satellite cities: Tokyo's 20th century experience of planned polycentrism. International Journal of Planning Studies 6(1):9–32.

Sorensen A (2002) The making of urban Japan: Cities and planning from edo to the 21st century. Routledge, London.

Tachibanaki T (1992) Higher land prices as a cause of increasing inequality: Changes in wealth distribution and socio-economic effects. In: Haley JO, Yamamura K (eds) Land issues in Japan: A policy failure? Society for Japanese Studies, Seattle, pp 175–194.

Ui J (ed) (1992) Industrial pollution in Japan. United Nations University Press, Tokyo.

Upham FK (1987) Law and social change in postwar Japan. Harvard University Press, Cambridge.

Watanabe Y (1992) The new phase of Japan's land, housing, and pollution problems. Japanese Economic Studies 20(4):30–68.

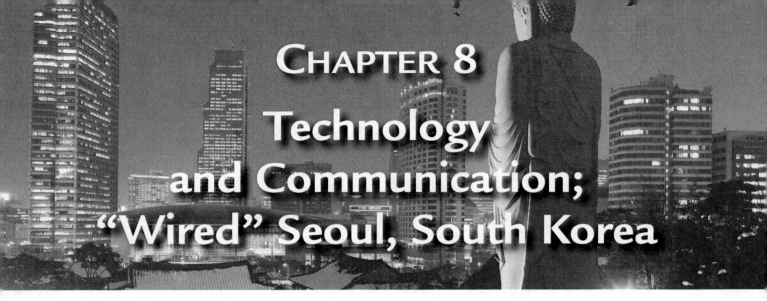

CHAPTER 8
Technology and Communication; "Wired" Seoul, South Korea

Seoul: World's Most Wired Megacity Gets More So

The world's most wired metro area is extending city services with its U-city project. The *U* stands for ubiquitous

Bill Powell and Stephen Kim

In the sprawling, densely populated capital city of South Korea, Lee Hye-young and her husband Kim Soon-kyo are nothing if not typical citizens. Which is to say, even the most mundane, everyday aspects of their lives are carried out at technology's leading edge.

Consider their respective commutes to work early one recent morning. Lee clambers onto a city bus, headed to her office job in the southern part of the city. She pays using her radio-frequency-identification (RFID) card—it has a computer chip in it—part of a transit program conceived and implemented by the city government. The card is smart enough to calculate the distance she travels on any form of public transit, which determines the fare. She can then use the same card to pay for the taxi she hails to finish her journey to work. Sometimes her husband, the deputy marketing manager at a small chemical company, drives her to work. But not today. A few months ago, he applied online to join a program offered by the city that promises insurance discounts, reduced-cost parking and a tax break if he leaves his car home one business day a week. The city sent him an RFID tag, which he attaches to the windshield so the city can monitor compliance. It took him just minutes to fill out the application on his home computer, and now, he says, he saves the equivalent of $50 a month. From the city's standpoint, the estimated 10,000 fewer cars on the road each day means less congestion and less air pollution in one of the busiest cities in East Asia.

For a decade, Seoul has had the justifiable reputation of being one of the most wired cities in the world. After the Asian financial crisis devastated the South Korean economy in 1997, the Seoul city government, the national government and the private sector all made a concerted effort to move the country's economy from one reliant on heavy industry to one that included information technology—a shift that by most measures has been a resounding success. Today, according to data compiled by Strategy Analytics, a U.S.-based technology market-research firm, an astonishing 95% of households in South Korea have a broadband connection. (Tiny Singapore is second, at 88%, and the U.S. comes in at No. 20, with just 60% hooked to broadband.) The entire city of Seoul, whose metro-area population is more than 20 million, is already one giant hot spot, with wireless access available from virtually anywhere within city limits for a small fee.

That level of connectedness, either via high-speed cable or through the ether, has not only transformed South Korea's economy; it has changed forever the way this massive city is governed, how individuals receive services and interact with city hall and how prospective contractors solicit business with the city.

Start with clean government. All city contracts are now put out to bid online, and all bids are posted. That transparency, Seoul Mayor Oh Se-hoon tells TIME, has reduced corruption in the city significantly in the past 10 years. "Since all information is disclosed real time over the Internet, influence-peddling over the bargaining of government permits becomes impossible," he says. "The online system tracks the flow of approval routes and leaves behind evidence in real time. If a manager holds on to an application for too long, he becomes a suspect. So administration becomes faster and uncorrupt." And while every big-city mayor may boast that his government is less corrupt than the last guy's—and corporate corruption has been an acknowledged problem in South Korea—Seoul has been named the world's most "advanced and efficient e-government" for several years by a U.N.-sponsored e-government-evaluation agency.

The city services accessible via Internet technology are already vast and growing rapidly. When Lee was returning home from work one day, she needed to pick up a copy of her social-security certificate. She did so at a subway station near her office, using a fingerprint-recognition kiosk: she placed her thumb on the machine, it read her print, and out popped a copy of the document. If she had so desired, she could have also printed real estate and vehicle registrations. It goes without saying that Lee pays her city taxes and utility bills online—or with her mobile phone's browser—and recently she dialed 120 to find out why the electric company had overcharged her. She was calling the Dasan Call Center, a 24/7 government agency that fields all questions regarding city services. A service rep did a quick check, confirmed the error and made sure her bill for the next month would reflect the correction.

Seoul has even greater e-ambitions. It has begun to implement a project called Ubiquitous Seoul— or U-city—which will extend the city's technological reach. Seoul's nearly 4-mile-long (6 km) Cheonggye Stream walkway, which runs through the high-rises of downtown Seoul, is the site of a U-city pilot project. Via their phones and laptops or on touchscreens located in parks and public plazas, citizens can check air-quality or traffic conditions or even reserve a soccer field in a public park. The city also sends out customized text messages. The city's chief information officer, Song Jung-hee, says those with respiratory problems can get ozone and air-pollution alerts, and commuters can get information about which route is the most congested at any given time. The city calls these real-time, location-based services.

Earlier this year, the city rolled out U–safety zones for children, a program using security cameras, a geographic-information-system platform and parents' cell-phone numbers. Participating families equip their kids with a U-tag—an electronic signature applied to a coat or backpack that allows a child to be tracked at all times. If the child leaves a designated ubiquitous-sensor zone near a school or playground, an alarm is automatically triggered alerting parents and the police. The child is then located via his or her mobile phone. The city plans to increase such zones rapidly. To some Americans, the Big Brother–ish qualities of the U-city push can be a tad unnerving. But Seoul officials point out that the U-safety-zone project is entirely voluntary, and the technologically sophisticated citizens seem to have few objections.

Seoul over the past decade has become a hotbed of early adopters, and global powerhouses from Microsoft to Cisco Systems to Nokia use it as a laboratory. The level of connectivity provided by the city's electronic infrastructure means "ubiquitous life" has become an inescapable catchphrase in Seoul. "Almost all new apartment complexes now advertise home networks and ubiquitous-life features," says Lim Jin-hwan, vice president for solution sales at Samsung Electronics. In a nutshell, that means every electronic device in the home can be controlled from a central keypad or a cell phone. Biorecognition lock systems open apartment doors, and soon, Lim says, facial-recognition systems will be introduced.

As megacities continue to grow and become more complex, it's likely that many will have to get wired just to stay manageable. Seoul took the considerable risk of being out front, but it has demonstrated the potential payback when the city government, and not just the citizens, is one of the early adopters.

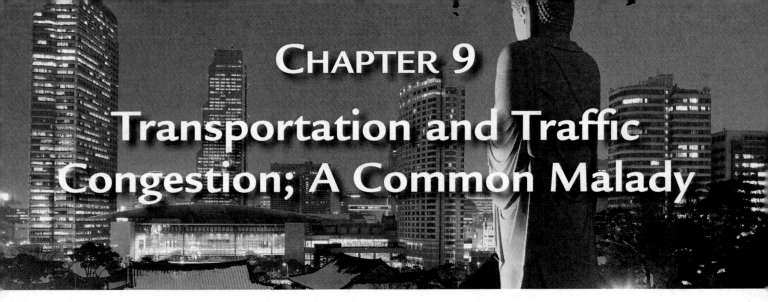

CHAPTER 9
Transportation and Traffic Congestion; A Common Malady

Urban Transport

INTRODUCTION

The provision of urban transport affects and is affected by a range of social, cultural, economic, political and environmental factors. In recent times, countries in the ESCAP region have undergone considerable change across all of these dimensions. For many of the region's urban areas, this change has come in the form of increased growth, in terms of population, economy and geographic size. That the region is now home to twelve of the world's mega-cities is evidence of this. Urban areas are expected to absorb a significant majority of future population growth in the region.

Urban centres are important hubs of economic activity. These areas, especially large cities, in general make a contribution to the national economy that is well in excess of their share of national population. For example, approximately 37.4 per cent of Thailand's economic activity takes place in Bangkok, which is home to 10.9 per cent of the Thai population.[1] It is understandable then that increasing urban income levels have also been part of the story of urban change in the region over recent years. Yet while the region's cities have made a major contribution to the growth in per capita income in the region, increasing urban poverty has also formed part of this story. Illustrating this, Asia still has the largest slum population in the world.[2]

Urban growth, in its many dimensions, creates increasing demand for effective urban transport systems. In many cities of the ESCAP region, however, the provision of urban transport has failed to keep up with the ever-increasing demand. In many cities, densification and spatial expansion have occurred with little or no development planning, while in some cases the failure of the instruments of governance has resulted in a significant wastage of resources or substandard quality of infrastructure. Furthermore, the huge capital costs and time required to develop high capacity transit systems have prevented the timely implementation of such systems in rapidly growing urban areas. As a result, many cities have relied on road-based systems, which have serious capacity constraints, negative environmental consequences and other limitations.

From "Urban Transport," *Review of Developments in Transport in Asia and the Pacific 2005*, Part 2, XI, Urban Transport, pp. 121–132. United Nations Transport Division, Economic and Social Commission for Asia & the Pacific (ESCAP). Copyright © United Nations 2005. Used by permission.
[1] UNHABITAT, 2004. *State of the World's Cities 2004/2005: Globalization and Urban Culture*, (Earthscan, London).
[2] Ibid.

GROWING MOTORIZATION

Cities in the ESCAP region do not have high motorization rates compared with others in the developed world, such as the European Union or the United States. With the exception of a few cities in Central Asia, however, in recent years the number of private vehicles in the region's cities has grown substantially. At the same time, the level and rate of motorization varies greatly between cities, due partly to differences in income levels and partly to government policy.

In 1990, there were 46 registered motor vehicles per 1,000 persons in Thailand. By 2001, this had increased to 359 per 1,000 persons.[3] Over this period, the number of motor vehicles in Bangkok grew by approximately 300,000 vehicles per annum.[4]

The vehicle population of Busan also grew by 22 per cent per annum between 1989 and 2001, at which point it totalled 862,699.[5] Some of the most rapid increases in motorization have taken place in China and India however.[6] For example, Mumbai has registered an annual growth of motorized vehicles of about 10 per cent in recent years, while between 1995 and 2000, Delhi's total motor vehicle population grew from 2.4 to 3.3 million, of which the car population increased from 576,000 to 837,000.[7] Meanwhile, Dhaka has maintained low motorization, with two-wheelers making up the greater proportional share of the total vehicle population.

One of the remarkable aspects of increased motorization in Asia, particularly in the region's mega-cities, is the increase in the number of two-wheelers. Thus, while relative to other countries, Dhaka has a very low motorization rate, the importance of two-wheelers to its total vehicle population, is a trend consistent with a number of other Asian mega-cities. According to data produced by the World Bank, for example, in 2001 India had 6 passenger cars per 1,000 people, compared with 29 two-wheelers. Two-wheelers, motorcycles in particular, are also the dominant transport mode in Hanoi: the city's one million registered motorcycles account for 60 per cent of all vehicle trips.[8]

While increases in motorization are related to growing income levels, they also contribute to traffic congestion and noise and air pollution. For example, transport accounts for up to 80 to 90 per cent of commercial energy consumption in Asia.[9] In addition, the unmanaged growth of motorization is the root cause of many of today's urban transport problems. Due to imperfect systems of transport pricing, prices do not reflect the true cost of provision of the transport services and facilities. Consequently, this has led to a waste of resources, insufficient funds to develop and maintain infrastructure, distortions in modal choice and the generation of externalities, such as pollution and congestion.

There are currently two ongoing regional intergovernmental air quality management programmes in the ESCAP region: the Air Pollution in the Megacities of Asia (APMA) project and the Clean Air Initiative for Asian Cities (CAI-Asia) project. In December 2000, the UNEP and WHO alongside the Korea Environment Institute and the Stockholm Environment Institute (SEI), with funding from the Korean Ministry of Finance and the Swedish International Cooperation and Development Agency (SIDA) initiated the APMA project. The World Bank and the ADB established the CAI-Asia project, in 2001. Research prepared for these projects shows growing motorization and the growth of the transport sector in general as a key driver of air pollution in Asia's mega-cities. At the same time, it points to the differing impacts and responses between these cities.

---

[3] ASEAN (2004), ASEAN 2003 Yearbook.

[4] Clean Air Initiative Asia web site, accessed August 2005, <http://www.cleanairnet.org>.

[5] Haq, G., Han, W., Kim, C., and Vallack, H. (2001) *Benchmarking Urban Air Quality Management and Practice in Major and Mega Cities of Asia: Stage I*, Report prepared and published under the APMA Project.

[6] The World Bank "Motorization, Demand & City Development", online document, accessed <http://www.worldbank.org/WBSITE/EXTERNAL/TOPICS/>, 11 August 2005.

[7] N.V. Iyer, *Measures to control vehicle population: The Delhi experience*, paper presented at the workshop on Fighting Urban Air Pollution: From Plan to Action, held at Bangkok from 12 to 14 February 2001.

[8] World Bank (2004), *VN-Hanoi Urban Transport Project*, [Project Information Document (PID)], (The World Bank, Washington), accessed via World Bank projects database, <http://www.worldbank.org>.

[9] UNCHS (HABITAT) (2001) "The Role of Urban Transport in Sustainable Human Settlements Development", *Background Paper No. 7 for Commission on Sustainable Development Ninth Session*, New York, 16–27 April 2001.

In Bangkok for instance, the transport sector is responsible for the emission of the majority of air-borne pollutants. This is linked to high incidences of cardiovascular and respiratory diseases, particularly in populations living or working near busy roads. Bangkok residents also have a high number of cases of throat irritation, at approximately 60 per cent of the population. In addition, increases in motorization have surpassed road network growth, producing traffic congestion, transportation delays and added pollution.

The high and increasing demand for transport in the way of highly polluting two-strokes, including auto-rickshaws, auto-tempos and motorcycles is also related to a variety of illnesses. Dhaka's air pollution is estimated to cause around 10,800 premature deaths and 6.5 million extra cases of sickness per annum, at a cost of between US$ 200-US$ 800 million.[10]

The costs of pollution and congestion in urban Asia are not only health and economic, but impact on the total quality of life of the urban population including the loss of leisure time because of excessive journey-to-work times. For example, data produced by the World Bank puts the estimated travel time to work in Bangkok and Seoul at 60 minutes.[11]

Most countries have addressed the challenge of growing motorization through a combination of increased investment in road stock and the development of complimentary public transport initiatives.

PUBLIC TRANSPORT INITIATIVES

Public transport has a very important role in urban transportation. In a number of cities in the region such as Hong Kong, China; Singapore and Tokyo, the modal share of public transport is 70 per cent or more of total person trips. In Central Asian cities, a high percentage of work trips are made using public transportation. In Bishkek for instance, work trips using public transport account for a 95 per cent share of the total. As has been discussed, the vast majority of the region's urban poor are heavily dependent on public transportation.

Compared with private cars, public transportation is more sustainable on economic, financial, social and environmental grounds. The failings of public transportation have become, however, one of the major challenges faced by many cities. Dissatisfaction with the level and quality of public transportation leads those people who can afford it to turn to private modes of transport.

In recent years, governments in the region's urban areas, particularly the large cities, have invested in Mass Rapid Transit (MRT) projects. The term "Mass Rapid Transit" refers to "public transport modes operating on fully or partially exclusive tracks (rail or road), away from street traffic and thus subject to full or at least considerable managerial control by the operator".[12] In a number of cases, these projects are undertaken with private sector participation. MRT systems have many advantages. Most notably, MRT systems can provide high capacity as well as high quality services. At the same time, however, they require considerable funding, their value contingent upon high passenger flows. Concomitantly, MRT systems tend to suit and feature in large and densely populated cities.

Urban Rail Projects

Governments of many countries have begun studying or implementing projects to develop rail-based mass transit systems in response to the shortcomings of road-based transport systems to meet growing demand in very large cities. Several Asian cities have addressed this issue through the implementation of rail-based mass transit systems. Underground rail systems are a long-established feature of Asia metropolises such as Tokyo; Hong Kong, China and Seoul. These systems continue to be enhanced and improved.

---

[10] Haq, Gary; Han, Wha-Jin and Kim, Christine (eds), 2002. *Urban Air Pollution Management in Major and Mega Cities of Asia*, (Korea Environment Institute, Seoul). Report prepared and published in the framework of the APMA project and is based on proceedings from the APMA workshop held in Seoul, Republic of Korea, 3–5 September 2001. Report accessed August 2005, <http://www.asiarnet.org>.

[11] World Bank, 2004. *World Development Indicators 2004* (World Bank, Washington).

[12] World Bank *Public Transport Modes & Services*, available <http//:w\www.worldbank.org>. Accessed 11 August 2005.

A number of urban rail projects are currently underway in China. In Sichuan Province in the country's south-west, a Metro is planned for the city of Chengdu. Pending approval from the State Council, construction of Line 1 of the Metro is scheduled for completion in 2008. When ready for operation, Line 1 will traverse 26 km from the zoo in the city's north, to Huayang in the south. Just over 11.7 km of the rail line will be underground. A planned second line will connect Honghe and Chengkwun, covering a distance of approximately 29 km.

Beijing is also investing in a metro network, including subways, light rail and suburban trains, in anticipation of the 2008 Olympics. In December 2002, construction began on the 27.6 km Subway Line 5. A north-south line, Subway Line 5 will have 16.9 km of track and 16 stations underground, with another 10.7 km and 6 stations above ground. Construction of Subway Line 4 and Line 10 began in December 2003 and are due for completion by December 2007. The former extends 28.6 km and the latter 30.5 km.[13]

The Beijing Light Rail is another important component of the programme of investments in infrastructure made in preparation for the 2008 Olympics. The elevated Light Rail network connects the eastern suburbs of Tongzhou to the subway terminus at Sihuidong. Construction on the Light Rail began in 2000. Line 13, from Xizhimen to Huoying opened in September 2002, while the remaining network opened for operation in 2003. At June 2005, however, Beijing is reported to have delayed the construction of the Light Rail link connecting the airport and the city centre, a project valued at Rmb 5 billion (US$ 604 million). Construction on the link, which was due to open in 2007, was delayed as the approval of the National Development and Reform Commission was still pending.[14]

Shanghai is reported to have 11 metro lines, totalling more than 300 km, and 10 light rail lines, totalling 120 km, planned for construction over the next 25 years.[15]

In India, expansion of the Delhi Metro system is currently underway, as part of the Delhi Metro master plan. Scheduled for completion by 2021, the master plan includes the construction of 240 km of high capacity rail transit. It is expected that Phases I and II of the project will be completed by 2005 and 2010, respectively. Phase I includes 3 lines. Line 1 stretches 22 km between Shahdara, Tri Nagar and Rithala. Line 2 connects Vishwa Vidyalaya and the Central Secretariat, at a length of 11 km. Finally, the 32.1 km-long Line 3 extends from Indraprastha to Dwarka Sub City via Barakhamba Road.

Development of the Delhi Metro has been under consideration for over fifty years. Significant progress was made towards achieving this goal in 1995, with the registration of the Delhi Metro Rail Corporation (DMRC). Then in October 1998, construction of the Metro started. Between December 2002 and July 2005, a number of sections of Metro Lines 1 and 2 have been completed. The most recent of these was the 7-km Central Secretariat – Inter-State Bus Terminus (ISBT)/Kashmere Gate section. The DMRC has estimated that ridership on the Metro in 2005 to be 1,500,000 passenger trips per day.[16]

The DMRC is also developing commercial property around and within stations, with the involvement of the private sector. For example, the development of commercial property around Kashmere Gate is currently up for tender and is offered under a 12-year concession contract. Similarly, the development at Khayala, adjacent to the Subash Nagar station on the Barakhamba—Dwarka corridor (Line 3), includes the construction of commercial services such as retail shops, an amphitheatre and an office complex.[17]

The DMRC has proposed the construction of a three-corridor, 37 km elevated rail system for Hyderabad. In early 2005, Hyderabad's urban development authorities made an interim recommendation to approve the project, to be financed through a build-own-transfer scheme. According to a

[13] <http://www.urbanrail.net>. Accessed August 2005.

[14] Dickie, Mure. "Beijing delays work on Olympic rail link", *Financial Times*, 20 June 2005. Accessed via Financial Times web site, August 2005, <http://www.financialtimes.com> and Stewart, "Olympic link plan", *Railway Gazette International*, 1 December 2004.

[15] <http://www.urbanrail.net>. Accessed August 2005.

[16] Delhi Metro Rail Corporation web site, accessed August 2005, <http://www.delhimetrorail.com>.

[17] Delhi Metro Rail Corporation web site, accessed August 2005, <http://www.delhimetrorail.com> complemented by data from <http://www.urbanrail.net>.

pre-feasibility study for a BRT network for the city prepared by the Institute for Transportation and Development Policy (ITDP), the rail system would require $ 1.1 billion of capital investment, for which amount a 294-km BRT system could be built.[18]

In Indonesia, the construction of a monorail system for Jakarta began in 2004. The system consists of two lines, which DKI Jakarta previously identified as suitable for monorail. The green line is to have a total length of 14.8 km with 17 stations; the Blue Line will be 12.2 km-long and have 12 stations. Since construction began, the project has experienced some setbacks however. Most notably, PT Jakarta Monorail, the company responsible for the construction of the monorail, halted construction after requesting more than US$ 20 million annually in operating subsidies for seven years as well as government equity investment in order of US$ 60 million.[19]

In September 2005, Phase I of the Almaty Metro Municipal project is expected to get underway. The project has three phases in total, covering the construction of some 40 km of track and 44 stations. When operational, the system will be able to support 500,000 passengers per day. A 400-room hotel, and shopping mall as well as fast food services in each station are also part of the project.

In Manila, a Rail-Mass Transit system, in the form of a light metro system, is currently being developed.

Three different types of metros comprise Bangkok's mass transit network. The Skytrain was the first rail transit system in Bangkok has been in operation since December 1999. Also known as the Green Line, the Skytrain was built by Siemens and is currently operated by the Bangkok Mass Transit System (BMTS). Two lines, namely, the Sukhumvit and Silom lines comprise the 23 km elevated rail transit system, which service the inner city as well as many commercial, residential and tourist areas of Bangkok. An 8.9-km extension of the Sukhumvit line from On Nut Station to Samrong is currently under construction.

An underground subway is the second component of Bangkok's mass transit network. The subway was formerly known as the Blue Line, but has since been renamed the Chaloem Ratchamongkhon, meaning "Celebration of the Auspicious Kingship". Bangkok Metro operates the subway under a concession agreement. The government has approved a number of extension projects, including the 29 km Purple Line, from Bang Yai to Bang Sue. These are expected to be complete by 2009.

A 24-km elevated road and rail structure was also planned as the third component of the mass transit network. Due to financial difficulties, however, construction was suspended in 1998.

A Light Rail system is also planned for Thailand's northern city of Chiang Mai.

Other cities in which rail-based mass transit systems have either been implemented, are planned or are under active consideration include Busan and Incheon (Republic of Korea), Kolkata (India), Daegu and Tianjin (China), Bangalore, Karachi and Mumbai (India), Karachi (Pakistan) and Dhaka (Bangladesh).

Bus Rapid Transit Projects

Bus-based MRT, or Bus Rapid Transit (BRT) systems provide alternatives to other MRT, as BRT systems tend to be cheaper and faster to construct, more profitable to operate and cheaper for commuters. Due in part to the success of BRT projects in Latin American cities, in recent years a number of cities in the ESCAP region have undertaken similar initiatives. More specifically, BRT systems are either operational, planned, under construction or under consideration in 36 cities in 10 countries in Asia. A large proportion of Asia's operational BRT systems are in Japan; the majority of those that are planned, under construction or under consideration are in Chinese cities.[20] Table 9.1 provides an overview of a selection of these projects.

[18] Institute for Transportation and Development Policy, "Hyderabad, India Betting the Bank on Elevated Rail", Sustainable Transport e-update no.18, August 2005. Accessed via ITDP web site, August 2005, <http://www.itdp.org>.
[19] <http://www.urbanrail.net>. Accessed August 2005.
[20] Clean Air Initiative for Asian Cities (CAI-Asia) web site, August 2005, <http://www.cleanairasia.org>.

Table 9.1   Selected Bus Rapid Transit (BRT) Projects in the ESCAP Region

City	Initiative
Bangkok	The establishment of the Bangkok BRT system is part of a greater portfolio of mass transit projects. The Bangkok Metropolitan Administration (BMA), who also own Bangkok's Skytrain, completed plans for the first two routes of the Bangkok BRT system in early 2005. The first of these is the 19.5 km Ram Intra Km 8 Navamin-Kaset-Mor Chit route; the second is the 16.5 km route connecting Chong Nonsi to Rama III, Krung Thep bridge and Rachapruek Road. When construction of these routes is complete, the Bangkok BRT system will have 42 new stations (25 on the first route and 16 on the second), 2 depots and 64 new buses (35 on the first route and 29 on the second). Stations will be positioned at 700-m intervals along each route. Access to these stations will be via footbridges, equipped with ticket booths and gates similar to those on the BTS electric train system. Key features of the buses used on the BRT system include air conditioning, Euro III standard engines, three right-side doors, and a carrying capacity of 150 sitting and standing passengers. The construction of the system's infrastructure and the procurement of the buses is estimated to total 2.7 million baht and is due for completion in October 2005. Depending on the success of these first two routes, an additional nine routes, with a combined length of 380 km, may be added to the system.[21]
Beijing	The development of the Beijing BRT is part of the Energy Foundation's *China Sustainable Energy Program*, funded by the, Hewlett Packard and Blue Moon foundations. With its first stage opened for operation in December 2004, the Beijing BRT system is the second 'closed' BRT system outside Latin America. Totalling some 16 km punctuated by 18 stops along the city's main north-south axis, the cost of the project was $ 72 million.
New Delhi	In January 2004, the Delhi Government approved funding for completing the technical designs for seven corridors of a *High Capacity Bus System* (HCBS) and for 10 high-capacity low-floor buses. The Indian Institute of Technology's Transportation Research and Injury Prevention Program (IIT-TRIPP) originally proposed the system. In collaboration with Rail India Technical and Economic Services (RITES), an engineering firm, ITT-TRIPP have completed two of the seven designs. The proposed HSBC was to operate mainly using normal buses and as an 'open' busway. Therefore, the HSBC is not strictly a BRT system, as "it does not have closed, pre-paid boarding stations, did not involve a re-routing of existing bus routes to a trunk and feeder system, and does not involve shifting bus operators to a payment per km basis".[22] Nonetheless, the HCBS is to have a number of distinct features, including separate busways, special bus shelters, feeder services, and safe, high quality pedestrian and cycling facilities. Its design also provides for the integration of cycle rickshaw parking and vendor sites.[23] In August 2004, final funding for the construction of the first corridor was approved, with construction expected to begin in early 2005. At August 2005, however, construction had not begun.
Hanoi	With financial assistance from the World Bank, the Japan Policy and Human Resources Development Fund and the Global Environment Fund (GEF), the Hanoi People's Committee Transport and Urban Works Projects Management Unit is investing US$ 170 million in an ongoing *Urban Transport Development Project* (HUTDP). The project has three components. The first of these covers an estimated US$ 60-US$ 90 million of investment into developing the city's bus network. Project activities include increasing the capacity of the bus system, building two experimental BRT routes, developing bus maintenance facilities, and the implementation of modern secure ticketing systems. The project is due for completion in 2010.[24]

---

[21] Bangkok Post Reporters, "TRANSPORT/BUS RAPID TRANSIT PROJECT; "Nine more routes planned for city", *Bangkok Post*, 2005.

[22] Fjellstrom, Karl, Diaz, Oscar Edmundo, and Gauthier, Aimee. "BRT's Great Leap Forward", *Sustainable Transport*, (Winter, 2004): 24–28 accessed via <http://www.itdp.org>.

[23] The Transportation and Research and Injury Prevention Program (TRIPP) "High Capacity Bus Systems", *TRIPP Bulletin*, V. 1, N. 3, Winter 2004 accessed via TRIPP web site August 2005, <http://www.iitd.ac.in/tripp>.

[24] World Bank (2004), *VN-Hanoi Urban Transport Project*, [Project Information Document (PID) Concept Stage], (World Bank, Washington), accessed via World Bank projects database, <http://www.worldbank.org>.

City	Initiative
Jakarta	Opened in January 2004 by the municipal government of Jakarta, TransJakarta is the first full BRT to be built in Asia. The first line in a planned 14-line system, measures 12.9 km and operates along the North-South corridor in the city centre, connection Jakarta's Northern rail station to the Blok M Bus Terminal and shopping area. Two additional corridors are currently under construction, expected to open by the end of 2005. The first of these, the east-west corridor, is expected to be expanded westward to a large bus terminal, and eastward to a new terminal in 2006. By June 2005, daily ridership on the BRT system had almost doubled from a year previous, to close to 65,000 passengers. As the system expands further, however, it is anticipated that passenger numbers will continue to increase on the first line, producing overcrowding. It is reported that, in anticipation of this, the city will add additional doorways to bus shelters, and will introduce articulated buses with more doors to the system. These improvements are expected to halve the present boarding time and increase the capacity of the first corridor from 2,700 to 6,000 passengers per hour in each direction.[25]
Shanghai	A BRT network is currently under consideration for Shanghai. EMBARQ, the Washington-based World Resources Institute's Center for Transport and the Environment, will provide some financial assistance for the project. The proposed network is to complement Shanghai's metro network, and it is reported that 100 and 150 km of BRT network will be constructed over the next five years. When complete, these lines will service both downtown and suburban areas, absorbing 20 per cent of the city's daily commuter traffic.[26]

Public Transport System Integration

While rail and bus-based mass transit systems each have their advantages, their full potential can best be realized when both form part of an integrated urban public transport system. In recent years, a number of cities in the ESCAP region have sought, or begun to seek ways of integrating their various modes of public transportation. Cities with more advanced forms of transportation, such as Hong Kong, China and Singapore, have successfully integrated their different public transport services provided by various operators, such as the underground and bus systems. Seoul and Metro Manila, meanwhile, have been less successful in modal integration.

In Bangkok, plans are underway to integrate the city's public transportation services. July 2005, Prime Minister Thaksin Shinawatra approved plans to establish a holding company to oversee the integration of Bangkok's mass transit system, with view to integrate all mass-transit operators into a single operator. As a result, commuters will be able to move between transit systems with ease, using a single ticket.

The holding company is to be a joint venture between the Finance Ministry and the MRTA. Under this proposal, the Bangkok Metro was to issue the MRTA with approximately 2.6 billion Baht worth of shares. The realization of an integrated mass-transit system experienced some obstacles, however, following the refusal of the BTS to proceed with share-sale negotiations with the government.[27] Reportedly, the Transport Ministry, therefore, ordered a study of alternative mass-transit routes.

The integration of public transportation modes in New Delhi, India is also being planned. The DMRC has signed an agreement with the Delhi Transport Corporation (DTC), operator of the city's bus services, to integrate services. Intra- and intercity buses would connect with the metro at the Inter-State Bus Terminus (ISBT), known also as Kashmere Gate and Connaught Place. In July 2005, the

---

[25] "Jakarta Announces Improvements to BRT", accessed via Institute for Transport and Development web site <http://www.itdp.org> and Fjellstrom, Karl, Diaz, Oscar Edmundo, and Gauthier, Aimee. "BRT's Great Leap Forward", *Sustainable Transport*, (Winter, 2004): 24–28 accessed via ITDP web site, August 2005, <http://www.itdp.org>.

[26] Organization of Asia-Pacific News Agencies, "Shanghai to Build 150 km Bus Rapid Transit", *Industry Updates*, 25 March 2005.

[27] "Mass Transit: Alternative to new BTS routes eyed", *The Nation*, 20 July 2005, accessed via The Nation web site, August 2005, <http://www.nationmultimedia.com> and Jalimsin, Aranee, "Overhauled mass transit system for Bangkok to include rapid routes", *Bangkok Post*, 29 November 2004.

Delhi Cabinet agreed in principle to have a common ticketing system for DTC and Metro Rail services in place by the end of the 2005. In addition, a bus feeder service, "Metro Link" would be established. The Department of Transport is planning to introduce a fleet of 200 low-floor and semi-low floor air-conditioned buses to be used in the feeder and local services.[28]

PARA-TRANSIT

Public transport in many Asian cities is characterized by a mix of formal public transport routes—often publicly operated—and a wide range of both motorized and non-motorized conveyances available for hire to public. Less formal means of public transport play a particularly important role in low to middle income countries.

As addressed previously, compared with other cities Dhaka has a low motorization rate. In fact, automobiles make up only 9 per cent of Dhaka's traffic; non-motorized transport (NMT) modes, particularly cycle rickshaws and walking account for the majority of trips. Dhaka's total cycle rickshaw population has been estimated at 500,000. Women, school children are the members of Dhaka's population most dependent on cycle rickshaws.[29]

Buses and minibuses remain the main forms of motorized public transport in the city. According to the Dhaka Transport Coordination Board (DTCB), in 2004 the city had approximately 22,000 registered private buses and minibuses and 400 Bangladesh Road Transport Corporation (BRTC) buses. Dhaka also has in order of 9,500 taxis and 10,000 auto rickshaws. Calculating the exact number of rickshaws in the city is also difficult. The DTCB reports that a large number of rickshaws enter the city daily, increasing the total rickshaw fleet to up to 80,000 vehicles.[30]

Bangkok has a large paratransit fleet of 49,000 licensed taxis; 7,400 3-wheeler *tuk-tuks;* 8,400 silor-leks (small 4-wheelers), and about 40,000 hired motorcycles (which provide services in lanes off the main roads). The majority of taxis and all tuk-tuks are LPG-powered. A recent innovation was the introduction of mini vans by the informal sector. About 8,000 14-seater minivans serve commuters on 103 routes, mainly between suburban locations and the central areas.

NON-MOTORIZED TRANSPORT

Non-Motorized Transport (NMT) including walking, remains a viable option to meet the basic mobility needs of all groups in a viable way. In fact, between 40 to 60 per cent of all trips in several major cities in Asia are made using NMT.[31] Furthermore, NMT is also an environmentally sustainable form of transportation. It is somewhat surprising, therefore, that "NMT has tended to be ignored by policymakers in the formulation of infrastructure policy and positively discouraged as a service provider".[32]

A number of countries in the ESCAP region are promoting the use of the bicycle as a sustainable means of transportation. For example, the Environmental Planning Collaborative (EPC) is redesigning some 120 km of road space in Ahmedabad, the largest city in the Indian state of Gujarat, with view to create exclusive bicycle lanes.[33]

Bicycle lanes have also been developed in Marikina, Metro Manila. This development is a component of the greater Metro Manila Urban Transport Integration Project (MMURTRIP), co-financed by the World Bank and the Government of the Philippines. Implemented by Global Environment Facility

[28] Staff Reporter, "Metro, DTC to have common tickets soon", *The Hindu*, 16 July 2005.

[29] Institute for Transportation and Development Policy, "World Bank says Dhaka rickshaw ban should not go ahead", Sustainable Transport E-Update, no. 16, April 2005, accessed via Institute for Transportation and Development web site, August 2005, <http://www.itdp.org>.

[30] Ibid.

[31] The World Bank, 2002. *Cities on the Move: a World Bank Urban Transport Strategy Review*, (The World Bank, Washington).

[32] Ibid.

[33] Institute for Transportation and Development Policy, "Ahmedabad, India moves ahead with bike lanes and BRT", Sustainable Transport E-Update, no. 18, August 2005. Accessed via Institute for Transportation and Development Policy web site, August 2005, <http://www.itdp.com>.

(GEF), the component includes the construction, evaluation and promotion of the Marikina Bikeway System (MBS)—a 66-km-long network of trails and road lanes designed specifically for NMT, plus bicycling parking and traffic calming systems. It is hoped that the new NMT-friendly facilities will encourage the use of NMT modes, and connection with the public transport terminals will promote the combined use of NMT and train/bus for trips between Marikina and the rest of metropolitan Manila. Full construction of the Bikeway is due for completion by the end of 2005.

The Metro Manila Development Authority (MMDA) has also proposed the establishment of bicycle lanes, under a "Foot and Pedal Ways Project". To this end, the MMDA has established a technical committee charged with planning the project. It is anticipated that the ground-level foot and pedal lanes, which would connect Metro Manila's 17 cities and municipalities, will be wider than the standard sidewalk width.[34]

CASE STUDIES: TRANSPORTATION IN BEIJING, COLOMBO AND KUALA LUMPUR

Case studies on urban transport development in Beijing, Colombo and Kuala Lumpur are presented in this section. These three cities are not representative of the whole range of urban transportation situations in the ESCAP region; their selection was rather determined by data and information availability.

Bejing

Beijing, the capital of China, is one of the ancient famous cities of the world with a long history. It has an area of 16,800 square kilometres and a population of about 13,819,000. The city is served by a modern multi-modal urban transportation system extending in all directions. In 2003, the total road length of the city was about 5,500 km. It included the newly rebuilt roads of 110 km. The road density is 85.4 km/100 square kilometres. Six expressways connect the city to other parts of the country.

In June 2005, the whole city had 2.41 million vehicles. The city's car population is swelling. In 2000, more than 60 per cent of the vehicles were private cars, which comprised over 23 per cent of the urban traffic volume. The share of the car traffic is increasing. It is estimated that by 2010 the total population will reach 16 million and the total vehicle population number will be 4 million.

Beijing is served by a well developed public transport system. In 2003, there were 776 bus lines, 16,939 buses and trams in total. The total length of public transport running lines was 15,760 km's. The estimated annual passenger volume is 4.7 billion, 27 per cent of which is carried by buses. At present, Beijing possesses 63,000 taxis, among which 48,000 are ordinary saloon cars and other 5,000 are top-grade cars. Recently, Beijing has begun to replace old taxis with new ones to prepare for the coming of 2008 Olympic Games. About two thirds of all the taxis had been replaced by the end of 2004.

The city's oldest subway opened to the public in 1969. Since then Beijing has been gradually developing a wide network of urban rail system comprising of elevated and underground systems. At present, there are five metro lines with a total length of 114 km's in operation. Another three lines and a branch line with a length of about 87 km's are under construction and will be open to traffic before the Olympic Games in 2008. Besides new lines, Beijing's two oldest metro lines will also be renovated ahead of the Olympic Games. Train cars and tracks along these lines will be upgraded by 2007 at a cost of US$ 450 million.

In preparation for the 2008 Olympic Game and to meet other demands, investment in urban infrastructure development has received a 'priority' attention. A number large transport projects have been implemented or are under construction. For example, a new line of elevated light railways has

---

[34] Ruiz, JC Bello. "BoTC urges drivers to prepare vehicles for proposed bike lanes", *Manila Bulletin*, 18 August 2005.

been opened to traffic recently. The construction of four new metro lines is also in progress. Further, construction of several highways, expressways, urban motorways and road network are being accelerated. The fourth and the fifth city ring road have been opened to traffic. Twenty-two special roads have been constructed for the 2008 Olympic Games. By the end of 2003, the total length of expressways in the city reached 499 km's and the total length of the roads had reached 14,453 km's.

The whole city has adopted a modern intelligent transportation system framework, which is led by an urban traffic control system based on real-time traffic information. Advanced equipment and technologies have been introduced for operation and management of urban public transportation system, which involves traffic control, resources allocation, and network optimization. About 80 per cent buses and 70 per cent taxis are now using clean fuel. The Euro II emission control standard was adopted a number of years back, which will be replaced by Euro III standard in 2005 and the Euro IV standard is planned to be adopted by 2008. Most of the main roads within the downtown area have adopted advanced anti-noise measures. Facilities specially designed for pedestrians and walking streets now provide a pleasant environment.

To provide support for holding a successful Olympic Games in 2008 Beijing is planning to construct the "New Beijing Transport System" with the characteristics of high quality infrastructure, advanced operation and management and an integrated control system. The main targets of the "New Beijing Transport System" include:

- Traffic congestion is expected to be eased greatly and the average speed in rush hour on main roads will reach 20 km per hour;
- The urban road structure will be improved with 15 new urban trunk roads connected with high-speed ring roads will be completed. The capacity of the road network will be increased by 50 per cent;
- The new urban bus transportation system will be completed. The total mileage of metro lines in operation will reach 300 km. A new bus rapid bus transportation system (BRT) will be developed and the running mileage will reach about 60 km;
- The share of public transport will reach 60 per cent;
- Operating efficiency and service level of taxis will be improved. By 2010, the daily passenger volume by taxis will reach 2.2 million/day and the empty running rate will be below 30 per cent;
- There will be significant improvement in road safety. The death rate of traffic accident will reduce to below 6 people/10,000 vehicles;
- Walking and bicycling will be the preferred main modes for short distance travelling; 80 per cent of short distance travels within 20 minutes should be undertaken by bicycling and walking; and
- Use of clean fuel will be more widespread and 90 per cent of public transport vehicles will use clean fuel.

Colombo

Colombo is located on the West coast of Sri Lanka. The city has an estimated resident population of 642,000. It has also a very large floating population. The area surrounding the Colombo Municipal Council is approximately 37 square kilometres. This is divided into 33 municipal wards and 15 postal zones. The average population density as per the 2001 census was 172 persons per hectare. The city functions as the administrative, commercial and educational capital of Sri Lanka even though an attempt has made to shift the administrative functions to an adjacent municipality of Sri Jayewardenepura. Suburbs of Colombo are also densely populated.

Colombo has a predominantly road based transport system. It has a road network of 480 km's. There are nine major road corridors that enter Colombo city. These include four main transport corridors (A1, A2, A3, A4) converging towards the city centre at the Colombo Municipal Council limits. At present there is no single main road that connects all these road corridors within or outside the city limits. Because of this reason, a large number of vehicles with both origins and destinations outside Colombo city, unnecessarily enter into the city and add to the growing traffic congestion on the main traffic corridors. The only main road that connects most of these corridors is the existing Baseline Road that connects seven out of nine main corridors. In addition to the nine major corridors, there are a couple of other secondary roads that also enter into the city of Colombo.

There are three main railway lines that radiate from Colombo and branch off at regular intervals. However, except for 116 km double-track lines around Colombo, the rest of the network is single-track line. Sri Lanka Railway is responsible for the operation of the system. The main sea port of Sri Lanka is also situated in Colombo and the lone international airport of the country is located at about 23 km's to the north of Colombo city.

It is estimated that on an average day nearly 360,000 vehicles carrying about 1,000,000 passengers enter or leave the boundaries of Colombo city. The majority of these vehicles are cars, vans or jeeps (25 per cent). There are significant proportions of motor cycles (11 per cent) and three wheelers (13 per cent). Buses make up 9 per cent of city traffic and trucks make up 6 per cent. Buses carry nearly 60 per cent of the commuters, while railway carriers about another 10 per cent during peak periods. The rest is carried by other modes including private transport and taxis.

Vehicle ownership level in Colombo is about 1 in 8 and 20 per cent of the registered motor vehicles of the country operate in Colombo. Nineteen per cent of all traffic accidents in the country take place in Colombo.

Public transport services in the city are provided by private and government-owned bus companies and by Sri Lanka Railways. Bus operation within Colombo is managed by the Western Provincial Council. Though these bus services are regulated to some extent, on-street competition and over loading is clearly visible. Buses do not follow any proper schedule at present. In addition to buses and trains, unregulated three-wheeler and demand responsive taxi services are in operation.

Except very few overpasses, all road intersections and road/rail crossings are at grade. Some of the busy intersections are controlled by traffic signals. A few signalized and un-signalized roundabouts are also present within the Colombo city limits. Traffic control is conducted by the Police Department. Most of the city roads experience traffic congestion especially during peak periods. On-street parking, other bottlenecks and mixed traffic conditions are the main reasons of traffic congestion.

Freight vehicles are restricted in certain parts of Colombo due mainly to security reasons. As a result, a number of truck routes have emerged for port bound freight vehicles. Even though axle load restrictions are in operation they are not strictly enforced. Overloading of freight vehicles is a very common occurrence in Colombo.

Colombo had been served by a canal transport system during the seventh to nineteenth centuries under the Dutch and British period of ruling. Today these waterways are hardly used for passenger transportation. Studies are underway to explore the possibility of reinstating canal based passenger transport services.

Kuala Lumpur

Kuala Lumpur is one of the largest cities in the region. It has an area of 243 square km's that houses administrative, commercial, industrial, educational and recreational activities and has all modern urban facilities and services to support these activities. The population of Kuala Lumpur was 1.4 million in 2000 and is expected to increase to 2.2 million over the next 20 years. In response to the ever growing demand for transport services, Kuala Lumpur has developed an urban transport system comprising an advanced public transport system and a modern road infrastructure network.

Responsibilities for urban transportation in Kuala Lumpur are divided between Kuala Lumpur City Hall and a number of ministries, department and agencies. The government is also considering a proposal for the creation of a Klang Valley Urban Transport Authority for the development and management of an integrated urban transportation system in the metropolitan area. The proposed authority will be responsible for public transport ownership, restructuring of regional, stage and feeder bus routes, new infrastructure development, fare structure design, electronic ticketing system, application of intelligent transport systems, maintaining a database and for research and development.

There has been dramatic increase in the level of motorization in recent years. From mere 85,000 vehicles registered in 1980, that number jumped to more than 2 million in 2000. In 2003, 58.1 per cent of the total registered vehicles in Kuala Lumpur were private cars. Motorcycles occupy about 30 per cent of total vehicles. The rapid growth of car-ownership has greatly changed the nature of demand for transport services. The modal share of public transport came down from 34.3 per cent in 1985 to 16.9 per cent in 2002. This increasing reliance on private cars has created considerable pressure on the roads and created congestion.

The government has taken a number of initiatives for the development of a modern transportation system for Kuala Lumpur. These include the development of integrated rail and bus services, electronic ticketing system, interchange facilities for smooth transfer between different modes and application of intelligent transport system for the management of road traffic and providing real-time information to road users.

The city is served by a well-developed public transport system with bus and four rail systems. KTMB runs two commuter lines with 39 stations. There are two (light) rail systems—PUTRA and STAR and also a short monorail system. The two rail systems totalling 56 km's of network are currently operated by a single operator, Rapid KL. The 8.6 km-long monorail system serves the inner city areas and provides links between the other rail systems. Four major bus operators provide 15,000 bus services per day operating on more than 30 routes. Feeder bus services complement the two light rail systems.

An impressive road and highway infrastructure development has taken place in recent years. The private sector has actively participated in the construction of roads and highways. For the smooth management of road traffic, City Hall has implemented an integrated transport information system. The ITIS has helped to improve traffic flow in the city. The system monitors traffic situation and congestion levels, automatically detects incidents and vehicle locations. It also updates real-time road condition messages on variable message sign boards along major roads.

In the 'Kuala Lumpur Structure Plan 2020,' emphasis has been placed into increasing the modal share of public transport. Various policy and development measures have been considered in the plan for the promotion and extension of public transport services in the city region. Some important measures include better integration of bus services with the rail systems via new multi-modal interchange facilities, and the establishment of transit planning zones in order to facilitate intensification of transit oriented residential, commercial and mixed-use development around rail stations. There are also efforts to regulate the parking facilities especially for freight transport and coaches in suitable locations around the periphery of city centre.

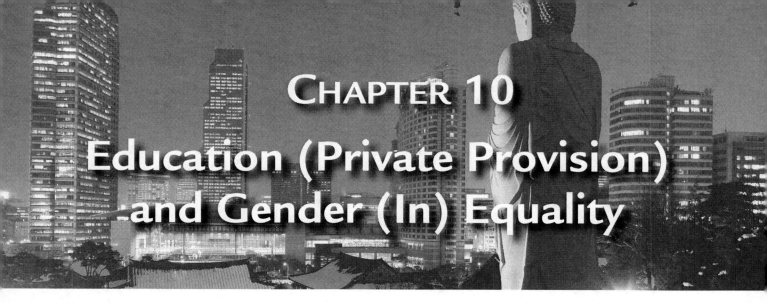

CHAPTER 10
Education (Private Provision) and Gender (In) Equality

The Many Faces of Gender Inequality

When Misogyny Becomes a Health Problem

Amartya Sen

It was more than a century ago, in 1870, that Queen Victoria wrote to Sir Theodore Martin complaining about "this mad, wicked folly of 'Woman's Rights.'" The formidable empress certainly did not herself need any protection that the acknowledgment of women's rights might offer. Even at the age of eighty, in 1899, she could write to Arthur James Balfour that "we are not interested in the possibilities of defeat; they do not exist." Yet that is not the way most peoples lives go, reduced and defeated as they frequently are by adversities. And within every community, nationality, and class, the burden of hardship often falls disproportionately on women.

The afflicted world in which we live is characterized by a deeply unequal sharing of the burden of adversities between women and men. Gender inequality exists in most parts of the world, from Japan to Morocco, from Uzbekistan to the United States. Yet inequality between women and men is not everywhere the same. It can take many different forms. Gender inequality is not one homogeneous phenomenon, but a collection of disparate and inter-linked problems. I will discuss just a few of the varieties of the disparity between the genders.

Mortality inequality. In some regions in the world, inequality between women and men directly involves matters of life and death, and takes the brutal form of unusually high mortality rates for women and a consequent preponderance of men in the total population, as opposed to the preponderance of women found in societies with little or no gender bias in health care and nutrition. Mortality inequality has been observed and documented extensively in North Africa and in Asia, including China and South Asian nations.

Natality inequality. Given the preference for boys over girls that characterizes many male-dominated societies, gender inequality can manifest itself in the form of parents' wanting a baby to be a boy rather than a girl. There was a time when this could be no more than a wish—a daydream or a nightmare, depending on one's perspective. But with the availability of modern techniques to determine the gender of a fetus, sex-selective abortion has become common in many countries. It is especially prevalent in East Asia, in China and South Korea in particular; but it is found also in Singapore and Taiwan, and it is beginning to emerge as a statistically significant phenomenon in India and in other parts of South Asia as well. This is high-tech sexism.

Basic-facility inequality. Even when demographic characteristics do not show much anti-female bias or any at all, there are other ways in which women can get less than a square deal. Afghanistan may be the only country in the world where the government is keen on actively excluding girls from schooling (the Taliban regime combines this with other features of massive gender inequality); but there are many countries in Asia and Africa, and also in Latin America, where girls have far less opportunity for schooling than do boys. And there are other deficiencies in basic facilities available to women, varying from encouragement to cultivate one's natural talents to fair participation in social functions of the community.

Special-opportunity inequality. Even when there is relatively little difference in basic facilities including schooling, the opportunities for higher education may be far fewer for young women than for young men. Indeed, gender bias in higher education and professional training can be observed even in some of the richest countries in the world, in Europe and North America. Sometimes this type of asymmetry has been based on the superficially innocuous idea that the respective "provinces" of men and women are just different. This thesis has been championed in different forms over the centuries, and it has always enjoyed a great implicit, as well as explicit, following. It was presented with particular directness more than one hundred years before Queen Victoria's complaint about "woman's rights" by the Reverend James Fordyce in his *Sermons to Young Women* (1766), a book that, as Mary Wollstonecraft noted in *A Vindication of the Rights of Woman* (1792), had been "long made a part of woman's library." Fordyce warned the young women to whom his sermons were addressed against "those masculine women that would plead for your sharing any part of their province with us," identifying the province of men as including not only "war," but also "commerce, politics, exercises of strength and dexterity, abstract philosophy and all the abstruser sciences." Such clear-cut beliefs about the provinces of men and women are now rather rare, but the presence of extensive gender asymmetry can be seen in many areas of education, training, and professional work even in Europe and North America.

Professional inequality. In employment as well as promotion in work and occupation, women often face greater handicaps than men. A country such as Japan may be quite egalitarian in matters of demography or basic facilities, and even to a great extent in higher education, and yet progress to elevated levels of employment and occupation seems to be much more problematic for women than for men. In the English television series *Yes, Minister,* there was an episode in which the Minister, full of reforming zeal, is trying to ascertain from the immovable permanent secretary, Sir Humphrey, how many women are in senior positions in the British civil service. Sir Humphrey says that it is very difficult to give an exact number; it would require a lot of investigation. The Minister is insistent, and wants to know approximately how many women are in these senior positions. To which Sir Humphrey finally replies, "Approximately, none."

Ownership inequality. In many societies, the ownership of property can also be very unequal. Even basic assets such as homes and land may be very asymmetrically shared. The absence of claims to property can not only reduce the voice of women, it can also make it harder for women to enter and to flourish in commercial, economic, and even some social activities. Inequality in property ownership is quite widespread across the world, but its severity can vary with local rules. In India, for example, traditional inheritance laws were heavily weighed in favor of male children (until the legal reforms after independence), but the community of Nairs (a large caste in Kerala) has had matrilineal inheritance for a very long time.

Household inequality. Often there are fundamental inequalities in gender relations within the family or the household. This can take many different forms. Even in cases in which there are no overt signs of anti-female bias in, say, mortality rates, or male preference in births, or in education, or even in promotion to higher executive positions, family arrangements can be quite unequal in terms of sharing the burden of housework and child care. It is quite common in many societies to take for granted that men will naturally work outside the home, whereas women could do so if and only if they could combine such work with various inescapable and unequally shared household duties. This is sometimes called a "division of labor," though women could be forgiven for seeing it as an "accumulation of labor." The reach of this inequality includes not only unequal relations within the family, but also derivative inequalities in employment and recognition in the outside world. Also, the established persistence of this type of "division" or "accumulation" of labor can also have far-reaching effects on the knowledge and the understanding of different types of work in professional circles. In the 1970s, when I first started working on gender inequality, I remember being struck by the fact that the Handbook of Human Nutrition Requirements of the World Health Organization, in presenting "calorie requirements" for different categories of people, chose to classify household work as "sedentary activity," requiring very little deployment of energy. I was not able to determine precisely how this remarkable bit of information had been collected.

It is important to take note of the implications of the varieties of gender inequality. The variations entail that inequality between women and men cannot be confronted and overcome by one all-purpose remedy. Over time, moreover, the same country can move from one type of gender inequality to another. I shall presently argue that there is new evidence that India, my own country, is undergoing just such a transformation at this time. The different forms of gender inequality may also impose adversities on the lives of men and boys, in addition to those of women and girls. In understanding the different aspects of the evil of gender inequality, we have to look beyond the predicament of women and examine the problems created for men as well by the asymmetrical treatment of women. These causal connections can be very significant, and they can vary with the form of gender inequality. Finally, inequalities of different kinds can frequently nourish one another, and we have to be aware of their linkages.

In what follows, a substantial part of my empirical focus will be on two of the most elementary kinds of gender inequality: mortality inequality and natality inequality. I shall be concerned particularly with gender inequality in South Asia, the so-called Indian subcontinent. While I shall separate out the subcontinent for special attention, I must warn against the smugness of thinking that the United States and Western Europe are free from gender bias simply because some of the empirical generalizations that can be made about other regions of the world would not hold in the West. Given the many faces of gender inequality, much depends on which face we look at.

Consider the fact that India, along with Bangladesh, Pakistan, and Sri Lanka, has had female heads of government, which the United States and Japan have not yet had (and do not seem very likely to have in the immediate future, if I am any judge). Indeed, in the case of Bangladesh, where both the prime minister and the leader of the opposition are women, one might begin to wonder whether any man could soon rise to a leadership position there. To take another bit of anecdotal evidence against Western complacence in this matter: I had a vastly larger proportion of tenured women colleagues when I was a professor at Delhi University—as long ago as the 1960s—than I had in the 1990s at Harvard University or presently have at Trinity College, Cambridge. And another example, of a more personal kind: when I was searching, a few years ago, for an early formulation of the contrast between the instrumental importance of wealth and the intrinsic value of human life, I found such a view in the words of Maitreyee, a woman intellectual depicted in the Upanishads, which date from the eighth century B.C.E. The classic formulation of this distinction, of course, would come about four centuries later, in Aristotle's *Nicomachean Ethics;* but it is interesting that the first sharp formulation of the value of living should have come from a woman thinker in a society that has not yet—three thousand years later—been able to overcome the mortality differential between women and men. In the scale of mortality inequality, India is close to the bottom of the league in gender disparity, along with Pakistan and Bangladesh; and natality inequality is also beginning to rear its ugly head very firmly and very fast in the subcontinent in our own day.

In the bulk of the subcontinent, with only a few exceptions (such as Sri Lanka and the state of Kerala in India), female mortality rates are very significantly higher than what could be expected given the mortality patterns of men (in the respective age groups). This type of gender inequality need not entail any conscious homicide, and it would be a mistake to try to explain this large phenomenon by invoking the cases of female infanticide that are reported from China or India: those are truly dreadful events, but they are relatively rare. The mortality disadvantage of women works, rather, mainly through the widespread neglect of health, nutrition and other interests of women that influence their survival.

It is sometimes presumed that there are more women than men in the world, since such a preponderance is well known to be the case in Europe and North America, which have an average female-to-male ratio of 1.05 or so (that is, about 105 women to 100 men). Yet women do not outnumber men in the world as a whole. Indeed, there are only about 98 women per 100 men on the globe. This "shortfall" of women is most acute in Asia and North Africa. The number of females per 100 males in the total population is 97 in Egypt and Iran, 95 in Bangladesh and Turkey, 94 in China, 93 in India and Pakistan, and 84 in Saudi Arabia (though the last ratio is considerably reduced by the presence of male migrant workers from elsewhere in Asia).

It has been widely observed that given similar health care and nutrition, women tend typically to have lower age-specific mortality rates than men. Indeed, even female fetuses tend to have a lower probability of miscarriage than male fetuses. Everywhere in the world, more male babies are born than female babies (and an even higher proportion of male fetuses are conceived compared with female fetuses); but throughout their respective lives the proportion of males goes on falling as we move to higher and higher age groups, due to typically greater male mortality rates. The excess of females over males in the populations of Europe and North America comes about as a result of this greater survival chance of females in different age groups.

In many parts of the world, however, women receive less attention and health care than do men, and girls in particular often receive very much less support than boys. As a result of this gender bias, the mortality rates of females often exceed those of males in these countries. The concept of the "missing women" was devised to give some idea of the enormity of the phenomenon of women's adversity in mortality by focusing on the women who are simply not there, owing to mortality rates that are unusually high compared with male mortality rates. The basic idea is to find some rough and ready way to understand the quantitative difference between the actual number of women in these countries and the number of women that we could expect to see if the gender pattern of mortality were similar there to the patterns in other regions of the world that do not demonstrate a significant bias against women in health care and other attentions relevant for survival.

We may take the ratio of women to men in sub-Saharan Africa as the standard, since there is relatively little bias against women in health care, social status, and mortality rates there, even though the absolute numbers are quite dreadful for both men and women. When estimating the size of the phenomenon of "missing women" in the mid-1980s, I used the prevailing female-male ratio in sub-Saharan Africa, around 1.022, as the standard. For example, with India's female-male ratio of 0.93, there is a total difference of 9 percent (of the male population) between that ratio and the sub-Saharan standard used for comparison. In 1986, this yielded a figure of 37 million missing women. Using the same sub-Saharan standard, China had 44 million missing women; and it became evident that, for the world as a whole, the magnitude of the gender shortfall easily exceeded 100 million. Other standards and other methods may also be used: Ansley Coale and Stephan Klasen have arrived at somewhat different numbers, but invariably very large ones. (Klasen's total number is about 80 million missing women.) So gender bias in mortality takes an astonishingly heavy toll.

How can this be reversed? Some economic models have tended to relate the neglect of women to the lack of economic empowerment of women. Ester Boserup, an early feminist economist, in her classic book *Women's Role in Economic Development,* published in 1970, discussed how the status and the standing of women are enhanced by economic independence (such as gainful employment). Others have tried to link the neglect of girls to the higher economic returns for the family from boys compared with girls. I believe that the former line of reasoning, which takes fuller note of social considerations that take us beyond any hard-headed calculation of relative returns from rearing girls vis-à-vis boys, is broader and more promising; but no matter which interpretation is taken, women's gainful employment, especially in more rewarding occupations, clearly does play a role in improving the life prospects of women and girls. So, too, does women's literacy. And there are other factors that can be seen as adding to the standing and to the voice of women in family decisions.

The experience of the state of Kerala in India is instructive in this matter. Kerala provides a sharp contrast with many other parts of the country in having little or no gender bias in mortality. The life expectancy of Kerala women at birth is above 76 (compared with 70 for men), and even more remarkably, the female-male ratio of Kerala's population is 1.06 according to the 2001 census, much the same as Europe or North America. Kerala has a population of 30 million, so it is an example that involves a fair number of people. The causal variables related to women's empowerment can be seen as playing a role here, since Kerala has a very high level of women's literacy (nearly universal for the younger age groups), and also much more access for women to well-paid and well-respected jobs.

One of the other influences of women's empowerment, a decline in fertility, is also observed in Kerala, where the fertility rate has fallen very fast (much faster, incidentally, than in China, despite Chinese coercive measures in birth control). The fertility rate in Kerala is 1.7 (roughly interpretable as an average of 1.7 children per couple), and it is one of the lowest in the developing world—about the same as in Britain and in France, and much lower than in the United States. We can see in these observations the general influence of women's education and empowerment.

Yet we must also take note of other special features of Kerala as well, including female ownership of property for an influential part of the Hindu population (the Nairs); openness to, and interaction with, the outside world (Christians form about one-fifth of the population and have been in Kerala much longer—since the fourth century—than they have been in, say, Britain, not to mention the very old community of Jews in Kerala); and activist left-wing politics with a particularly egalitarian commitment, which has tended to focus strongly on issues of equity (not only between classes and castes, but also between women and men). While these influences may work in the same way as the impact of female education and employment in reducing mortality inequality, they can have different roles in dealing with other problems, particularly the problem of natality inequality.

The problem of gender bias in life and death has been much discussed, but there are other issues of gender inequality that are sorely in need of greater investigation. I will note four substantial phenomena that happen to be quite widely observed in South Asia.

There is, first, the problem of the undernourishment of girls as compared with boys. At the time of birth, girls are obviously no more nutritionally deprived than boys, but this situation changes as society's unequal treatment takes over from the non-discrimination of nature. There has been plenty of aggregative evidence on this for quite some time now; but it has been accompanied by some anthropological skepticism about the appropriateness of using aggregate statistics with pooled data from different regions to interpret the behavior of individual families. Still, there have also been more detailed and concretely local studies on this subject, and they confirm the picture that emerges on the basis of aggregate statistics. One case study from India, which I myself undertook in 1983 along with Sunil Sen-gupta, involved weighing every child in two large villages. The time pattern that emerged from this study, which concentrated particularly on weight-for-age as the chosen indicator of nutritional level for children under five, showed clearly how an initial neonatal condition of broad nutritional symmetry turns gradually into a situation of significant female disadvantage. The local investigations tend to confirm rather than contradict the picture that emerges from aggregate statistics.

In interpreting the causal process that leads to this female disadvantage, it is important to emphasize that the lower level of nourishment of girls may not relate directly to their being underfed as compared with boys. Often enough, the differences may arise more from the neglect of heath care of girls compared with what boys receive. Indeed, there is some direct information about comparative medical neglect of girls vis-à-vis boys in South Asia. When I studied, with Jocelyn Kynch, admissions data from two large public hospitals in Bombay, it was very striking to find clear evidence that the admitted girls were typically more ill than the boys, suggesting that a girl has to be more stricken and more ill before she is taken to the hospital. Undernourishment may well result from a greater incidence of illness, which can adversely affect both the absorption of nutrients and the performance of bodily functions.

There is, secondly, a high incidence of maternal undernourishment in South Asia. Indeed, in this part of the world, maternal undernutrition is much more common than in most other regions. Comparisons of body mass index (BMI), which is essentially a measure of weight for height, bring this out clearly enough, as do statistics of such consequential characteristics as the incidence of anemia.

Thirdly, there is the problem of the prevalence of low birth weight. In South Asia, as many as 21 percent of children are born clinically underweight (by accepted medical standards), more than in any other substantial region in the world. The predicament of being low in weight in childhood seems often enough to begin at birth in the case of South Asian children. In terms of weight for age, around 40 to 60 percent of the children in South Asia are undernourished, compared with 20 to 40 percent undernourishment even in sub-Saharan Africa. The children start deprived and stay deprived. Finally, there is also a high incidence of cardiovascular diseases. Generally, South Asia stands out as having more cardiovascular diseases than any other part of the Third World. Even when other countries, such as China, show a greater prevalence of the standard predisposing conditions to such illness, the subcontinental population seems to have more heart problems than these other countries.

It is not difficult to see that the first three of these problems are very likely connected causally. The neglect of the care of girls and women, and the underlying gender bias that their experience reflects, would tend to yield more maternal undernourishment; and this in turn would tend to yield more fetal deprivation and distress, and underweight babies, and child undernourishment. But what about the higher incidence of cardiovascular diseases among South Asian adults? In interpreting this phenomenon, we can draw on the pioneering work of a British medical team led by D. J. P. Barker. Based on English data, Barker has shown that low birth weight is closely associated with the higher incidence, many decades later, of several adult diseases, including hypertension, glucose intolerance, and other cardiovascular hazards.

The robustness of the statistical connections and the causal mechanisms involved in the retardation of intrauterine growth can be further investigated, but as matters stand the medical evidence that Barker has produced linking the two phenomena offers the possibility of proposing a causal relation between the different empirical observations of the harsh fate of girls and women in South Asia and the phenomenon of high incidence of cardiovascular diseases in South Asia. This strongly suggests a causal pattern that goes from the nutritional neglect of women to maternal undernourishment, and thence to fetal growth retardation and underweight babies, and thence to greater incidence of cardiovascular afflictions much later in adult life (along with the phenomenon of undernourished children in the shorter run). In sum: what begins as a neglect of the interests of women ends up causing adversities in the health and the survival of all, even at an advanced age.

These biological connections illustrate a more general point: gender inequality can hurt the interests of men as well as women. Indeed, men suffer far more from cardiovascular diseases than do women. Given the uniquely critical role of women in the reproductive process, it would be hard to imagine that the deprivation to which women are subjected would not have some adverse impact on the lives of all people—men as well as women, adults as well as children—who are "born of a woman," as the Book of Job says. It would appear that the extensive penalties of neglecting the welfare of women rebound on men with a vengeance.

But there are also other connections between the disadvantage of women and the general condition of society—non-biological connections—that operate through women's conscious agency. The expansion of women's capabilities not only enhances women's own freedom and well-being, it also has many other effects on the lives of all. An enhancement of women's active agency can contribute substantially to the lives of men as well as women, children as well as adults: many studies have demonstrated that the greater empowerment of women tends to reduce child neglect and mortality, to decrease fertility and overcrowding, and more generally to broaden social concern and care.

These examples can be supplemented by considering the functioning of women in other areas, including in the fields of economics and politics. Substantial linkages between women's agency and social achievements have been noted in many different countries. There is plenty of evidence that whenever social and economic arrangements depart from the standard practice of male ownership, women can seize business and economic initiative with much success. It is also clear that the result of women's participation in economic life is not merely to generate income for women, but also to provide many other social benefits that derive from their enhanced status and independence. The remarkable success of organizations such as the Grameen Bank and BRAC (Bangladesh Rural Advancement Committee) in Bangladesh is a good example of this, and there is some evidence that the high-profile presence of women in social and political life in that country has drawn substantial support from women's economic involvement and from a changed image of the role of women.

The Reverend Fordyces of the world may disapprove of "those masculine women" straying into men's "province," but the character of modern Bangladesh reflects in many different and salutary ways the increasing agency of women. The precipitate fall of the total fertility rate in Bangladesh from 6.1 to 3.0 in the course of two decades (perhaps the fastest such decline in the world) is clearly related to the changed economic and social roles of women, along with increases in family-planning facilities. There have also been cultural influences leading to a re-thinking of the nature of the family, as Alaka Basu and Sajeda Amin have shown recently in *Population and Development Review*. Changes can also be observed in parts of India where women's empowerment has expanded, with more literacy and greater economic and social involvements outside the home.

There is something to cheer in the developments that I have been discussing, and there is considerable evidence of a weakened hold of gender disparity in several fields in the subcontinent; but the news is not, alas, all good. There is also evidence of a movement in the contrary direction, at least with regard to natality inequality. This has been brought out sharply by the early results of the 2001 decennial national census in India, the results of which are still being tabulated and analyzed. Early results indicate that even though the overall female-male ratio has improved slightly for the country as a whole (with a corresponding reduction of the proportion of "missing women"), the female-male ratio for children has suffered a substantial decline. For India as a whole, the female-male ratio of the population under age six has fallen from 94.5 girls per 100 boys in 1991 to 92.7 girls per 100 boys in 2001. While there has been no such decline in some parts of the country (most notably Kerala), it has fallen very sharply in Punjab, Haryana, Gujarat, and Maharashtra, which are among the richer Indian states.

Taking together all the evidence that exists, it is clear that this change reflects not a rise in female child mortality, but a fall in female births vis-à-vis male births; and it is almost certainly connected with the increased availability and the greater use of gender determination of fetuses. Fearing that sex-selective abortion might occur in India, the Indian parliament some years ago banned the use of sex determination techniques for fetuses, except as a by-product of other necessary medical investigation. But it appears that the enforcement of this law has been comprehensively neglected. When questioned about the matter by Celia Dugger, the energetic correspondent of *The New York Times*, the police cited difficulties in achieving successful prosecution owing to the reluctance of mothers to give evidence of the use of such techniques.

I do not believe that this need be an insurmountable difficulty (other types of evidence can in fact be used for prosecution), but the reluctance of the mothers to give evidence brings out perhaps the most disturbing aspect of this natality inequality. I refer to the "son preference" that many Indian mothers themselves seem to harbor. This form of gender inequality cannot be removed, at least in the short run, by the enhancement of women's empowerment and agency, since that agency is itself an integral part of the cause of natality inequality.

Policy initiatives have to take adequate note of the fact that the pattern of gender inequality seems to be shifting in India, right at this time, from mortality inequality (the female life expectancy at birth has now become significantly higher than male life expectancy) to natality inequality. And, worse, there is clear evidence that the traditional routes of combating gender inequality, such as the use of public policy to influence female education and female economic participation, may not, on their own, serve as a path to the eradication of natality inequality. A sharp pointer in that direction comes from the countries in East Asia that have high levels of female education and economic participation.

Compared with the biologically common ratio across the world of 95 girls being born per 100 boys, Singapore and Taiwan have 92 girls, South Korea only 88, and China a mere 86—their achievements in female empowerment notwithstanding. In fact, South Korea's overall female-male ratio for children is also a meager 88 girls per 100 boys, and China's grim ratio is 85 girls per 100 boys. In comparison, the Indian ratio of 92.7 girls per 100 boys (though lower than its previous figure of 94.5) looks far less unfavorable.

Still, there are reasons for concern. For a start, these may be early days, and it has to be asked whether with the spread of sex-selective abortion India may catch up with—and perhaps even go beyond—Korea and China. Moreover, even now there are substantial variations within India, and the all-India average hides the fact that there are states in India where the female-male ratio for children is very much lower than the Indian average.

Even though sex-selective abortion is to some extent being used in most regions in India, there seems to be something of a social and cultural divide across India, splitting the country in two, in terms of the extent of the practice and the underlying bias against female children. Since more boys are born than girls everywhere in the world, even without sex-specific abortion, we can use as a classificatory benchmark the female-male ratio among children in advanced industrial countries. The female-male ratio among children for the zero-to-five age group is 94.8 in Germany, 95.0 in the United Kingdom, and 95.7 in the United States. And perhaps we can sensibly pick the German ratio of 94.8 as the cut-off point below which we should suspect anti-female intervention.

The use of this dividing line produces a remarkable geographical split in India. In the states in the north and the west, the female-male ratio of children is uniformly below the benchmark figure, led by Punjab, Haryana, Delhi, and Gujarat (with ratios between 79.3 and 87.8), and also including the states of Himachal Pradesh, Madhya Pradesh, Rajasthan, Uttar Pradesh, Maharashtra, Jammu and Kashmir, and Bihar. The states in the east and the south, by contrast, tend to have female-male ratios that are above the benchmark line of 94.8 girls per 100 boys, such as Kerala, Andhra Pradesh, West Bengal, and Assam (each between 96.3 and 96.6), and also including Orissa, Karnataka, and the northeastern states to the east of Bangladesh.

Aside from the tiny states of Dadra and Nagar Haveli (with less than 250,000 people), which have a high female-male ratio among children despite being in the west, the one substantial exception to this adjoining division is Tamil Nadu, where the female-male ratio is just below 94—higher than the ratio of any state in the deficit list, but still just below the cut-off line (94.8) used for the partitioning. But the astonishing finding is not that one particular state is a marginal misfit. It is that the vast majority of the Indian states fall firmly into two contiguous halves, classified broadly into the north and the west on one side and the south and the east on the other. Indeed, every state in the north and the west (with the slight exception of tiny Dadra and Nagar Haveli) has strictly lower female-male ratios of children than every state in the east and the south (even Tamil Nadu fits into this classification). This is quite remarkable.

The pattern of female-male ratio of children produces a much sharper regional classification than does the female-male ratio of mortality of children, even though the two are also strongly correlated. The female-male ratio in child mortality varies, at one end, from 0.91 in West Bengal and 0.93 in Kerala, in the eastern and southern group, to 1.30 at the other end, in Punjab, Haryana, and Uttar Pradesh (with high ratios also in Gujarat, Bihar, and Rajasthan), in the northern and western group.

The pattern of contrast does not have any obvious economic explanation. The states with anti-female bias include rich states (Punjab and Haryana) as well as poor states (Madhya Pradesh and Uttar Pradesh), fast-growing states (Gujarat and Maharashtra) as well as states that are growth failures (Bihar and Uttar Pradesh). Also, the incidence of sex-specific abortions cannot be explained by the availability of medical resources for determining the sex of the fetus: Kerala and West Bengal in the non-deficit list have at least as many medical facilities as do the deficit states of Madhya Pradesh, Haryana, or Rajasthan. If the provision for sex-selective abortion is infrequent in Kerala or West Bengal, it is because of a low demand for those specific services, rather than any great barrier on the side of supply.

This suggests that we must inquire beyond economic resources or material prosperity or GNP growth into broad cultural and social influences. There are a variety of influences to be considered here, and the linking of these demographic features with the subject matter of social anthropology and cultural studies would certainly be very much worth doing. There is also some possible connection with politics. It has been noted in other contexts that the states in the north and the west of India generally have given much more room to religion-based sectarian politics than has the east or the south, where religion-centered parties have had very little success. Of the 197 members of the present Indian parliament from the Bharatiya Janata Party (BJP) and Shiva Sena, which represent to a great extent the forces of Hindu nationalism, as many as 169 were elected from the north and the west. While it would be important to keep a close watch on the trend of sex-selective abortion everywhere in India, the fact that there are sharp divisions related to culture and politics may suggest lines of probing investigation as well as remedial action.

Gender inequality, then, has many distinct and dissimilar faces. In overcoming some of its worst manifestations, especially in mortality rates, the cultivation of women's empowerment and agency, through such means as women's education and gainful employment, has proved very effective. But in dealing with the new form of gender inequality, the injustice relating to natality, there is a need to go beyond the question of the agency of women and to look for a more critical assessment of received values. When antifemale bias in behavior (such as sex-specific abortion) reflects the hold of traditional masculinist values from which mothers themselves may not be immune, what is needed is not just freedom of action but also freedom of thought—the freedom to question and to scrutinize inherited beliefs and traditional priorities. Informed critical agency is important in combating inequality of every kind, and gender inequality is no exception.

Educational Outcomes

The Asia-Pacific region is home to more than two thirds of the world's adult illiterates. The situation is particularly serious in South and South-West Asia, where over 390 million adults lack basic skills in reading, writing and numeracy, while one in four children never make it to the final grade of primary school.

In 2007, according to the 2009 Education for All Global Monitoring Report, the Asia-Pacific region had 510 million illiterate people. Almost two thirds were female. The situation is particularly difficult in South and South-West Asia: at least one in three adults is illiterate in Afghanistan, Bangladesh, Bhutan, India, Nepal and Pakistan. However, some countries in this region have seen significant improvements. Bangladesh and Nepal, for example, between 1991 and 2007, improved their adult literacy rates from 35.3 to 53.5% and from 33 to 56.5%, respectively.

Of the 31 countries that had data between 2005 and 2007, nine had literacy rates below 80%—Bangladesh, Bhutan, Cambodia, India, Lao People's Democratic Republic, Nepal, Pakistan, Papua New Guinea and Vanuatu.

The subregions with the highest literacy rates—above 90%—are North and Central Asia and South-East Asia and East and North-East Asia. However, more intensive and dedicated efforts on provision of literacy programmes through better access to cover all remaining ones as they tend to be the most vulnerable and most difficult to reach.

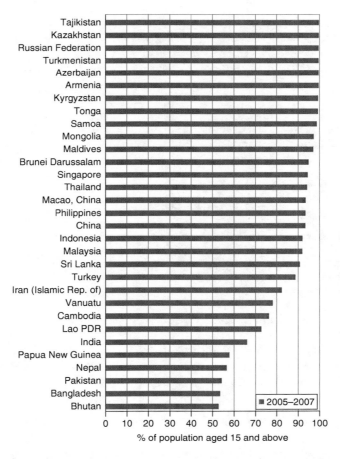

Figure 10.1   Adult literacy rates, Asia and the Pacific, 2005-2007.

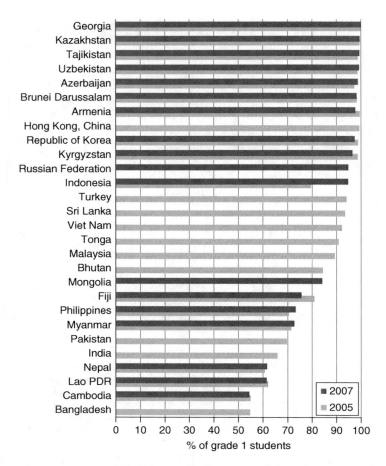

Figure 10.2   Survival rate to the last grade of primary level,
Asia and the Pacific, 2005–2007.

Survival Rate to the Last Grade of Primary Level (% of grade 1 students)

The proportion of pupils starting grade 1 who reach last grade of primary education is the percentage of a cohort of pupils enrolled in grade 1 of the primary level of education in a given school year who are expected to reach the last grade of primary school, regardless of repetition.

Source: United Nations Millennium Development Goals Indicators. Online database accessed on 31 August 2009.

Adult Literacy Rate (% of population aged 15 and above)

The proportion of literate persons among adult population, expressed as a percentage of population aged 15 years and above. Literacy is defined as the ability to read and write with understanding a simple statement related to one's daily life. It generally encompasses basic arithmetic skills. **Aggregates:** Calculated by UNESCO Institute for Statistics.

Source: UNESCO Institute for Statistics, Data Centre. Online database accessed on 14 August 2009.

Gender Parity Index for Adult Literacy Rate (ratio)

The number of literate women divided by the number of literate men in the population aged 15 years and above. **Aggregates:** Calculated by UNESCO Institute for Statistics.

Source: UNESCO Institute for Statistics, Data Centre. Online database accessed on 9 November 2009.

Table 10.1 Survival Rate to the Last Grade of Primary Level and Literacy

	Survival Rate to the Last Grade of Primary Level % of Grade 1 Students			Adult Literacy Rate % of Population Aged 15 and Above			Gender Parity Index for Adult Literacy Rate Ratio		
	2000	2005	2007	1990	2004	2007	1990	2004	2007
East and North-East Asia				**81.2**	**92.2**	**94.1**	**0.82**	**0.92**	**0.94**
China				77.8	90.9 (00)	93.3	0.78	0.91 (00)	0.93
DPR Korea									
Hong Kong, China		99.3 (04)							
Japan									
Macao, China					91.3 (01)	93.5 (06)		0.92 (01)	0.94 (06)
Mongolia	88.9		84.1 (06)		97.8 (00)	97.3		0.99 (00)	1.01
Republic of Korea	99.2	98.8	97.4 (06)						
South-East Asia				**85.1**	**90.2**	**91.4**	**0.89**	**0.94**	**0.94**
Brunei Darussalam		98.3	98.2 (06)	87.8 (91)	92.7 (01)	94.9	0.89 (91)	0.95 (01)	0.96
Cambodia	54.6	55.0	54.5 (06)	67.3 (98)	73.6	76.3	0.72 (98)	0.76	0.79
Indonesia	85.9 (01)	79.5	94.7 (06)	81.5	90.4	92.0 (06)	0.86	0.92	0.93 (06)
Lao PDR	53.2	62.0	61.5 (06)	60.3 (95)	68.7 (01)	72.7 (05)	0.65 (95)	0.79 (01)	0.77 (05)
Malaysia		89.3		82.9 (91)	88.7 (00)	91.9	0.87 (91)	0.93 (00)	0.95
Myanmar	55.2	71.5	72.7 (06)		89.9 (00)			0.92 (00)	
Philippines	75.3 (01)	70.4	73.2 (06)	93.6	92.6 (03)	93.4	0.99	1.02 (03)	1.01
Singapore				89.1	92.5 (00)	94.4	0.87	0.92 (00)	0.94
Thailand					92.6 (00)	94.1		0.95 (00)	0.97
Timor-Leste									
Viet Nam	85.7	92.1		87.6 (89)	90.3 (99)		0.89 (89)	0.93 (99)	
South and South-West Asia				**49.0**	**60.3**	**65.3**	**0.58**	**0.68**	**0.72**
Afghanistan					28.0 (00)			0.29 (00)	
Bangladesh		54.8		35.3 (91)	47.5 (01)	53.5	0.58 (91)	0.76 (01)	0.82
Bhutan	81.3	84.4				52.8 (05)			0.60 (05)
India	59.0	65.8		48.2 (91)	61.0 (01)	66.0	0.55 (91)	0.65 (01)	0.71
Iran (Islamic Rep. of)	97.5			73.1 (96)	77.0 (02)	82.3 (06)	0.83 (96)	0.84 (02)	0.88 (06)

(Continued)

Table 10.1 (*Continued*)

	Survival Rate to the Last Grade of Primary Level			Adult Literacy Rate			Gender Parity Index for Adult Literacy Rate		
	% of Grade 1 Students			% of Population Aged 15 and Above			Ratio		
	2000	2005	2007	1990	2004	2007	1990	2004	2007
Maldives	45.8			96.3 (95)	96.3 (00)	97.0	1.00 (95)	1.00 (00)	1.00
Nepal		60.8 (04)	61.6	33.0 (91)	48.6 (01)	56.5	0.35 (91)	0.56 (01)	0.62
Pakistan		69.7 (04)		42.7 (98)		54.2 (06)	0.52 (98)		0.58 (06)
Sri Lanka		93.4			90.7 (01)	90.8 (06)		0.97 (01)	0.96 (06)
Turkey		94.1 (04)		79.2	87.4	88.7	0.76	0.84	0.85
North and Central Asia				97.9	99.1	99.2	0.97	0.99	0.99
Armenia	79.3 (01)	99.5	97.7 (06)	98.8 (89)	99.4 (01)	99.5	0.99 (89)	0.99 (01)	1.00
Azerbaijan	96.9	97.3	98.7 (06)		98.8 (99)	99.5		0.99 (99)	0.99
Georgia	94.8	100.0	100.0 (06)						
Kazakhstan	95.9	99.4	99.5	97.5 (89)	99.5 (99)	99.6	0.97 (89)	0.99 (99)	1.00
Kyrgyzstan	93.0	98.6	96.5 (06)		98.7 (99)	99.3		0.99 (99)	1.00
Russian Federation	98.7		94.8 (06)	98.0 (89)	99.4 (02)	99.5	0.97 (89)	0.99 (02)	1.00
Tajikistan	95.5	98.7	99.4 (06)	97.7 (89)	99.5 (00)	99.6	0.98 (89)	0.99 (00)	1.00
Turkmenistan				98.8 (95)		99.5	0.99 (95)		1.00
Uzbekistan	97.8	98.6	99.2 (06)		96.9 (00)			0.98 (00)	
Pacific									
American Samoa									
Australia									
Cook Islands									
Fiji	86.1	81.0	75.6 (06)						
French Polynesia									
Guam									
Kiribati	69.7 (01)								
Marshall Islands									
Micronesia (F.S.)									
Nauru	25.4 (01)								

108

New Caledonia								
New Zealand	95.9							
Niue								
Northern Mariana Islands								
Palau								
Papua New Guinea				57.3 (00)	57.8		0.80 (00)	0.86
Samoa			97.9 (91)	98.6	98.7	0.99 (91)	0.99	0.99
Solomon Islands		90.9						
Tonga	94.6		98.9 (96)		99.2	1.00 (96)		1.00
Tuvalu	72.7							
Vanuatu	68.9 (99)		65.5 (94)	75.5	78.1	0.91 (94)	0.94	0.95
Asia and the Pacific			70.3	79.7	82.3	0.78	0.85	0.87
LLDC			74.9	76.6	77.5	0.86	0.87	0.87
LDC			47.0	56.7	60.4	0.67	0.78	0.81
ASEAN			85.1	90.3	91.5	0.89	0.94	0.94
ECO			64.4	69.0	72.9	0.75	0.79	0.80
SAARC			45.9	57.4	62.9	0.54	0.65	0.70
Central Asia			97.8	98.4	98.6	0.98	0.99	0.99
Pacific island dev. econ.			75.8	76.2	76.8	0.88	0.91	0.93
Low income			58.3	63.5	67.7	0.76	0.81	0.81
Middle income			71.2	81.6	84.3	0.77	0.85	0.88
High income			99.1	99.1	99.1	1.00	1.00	1.00
Africa			52.8	59.7	63.3	0.69	0.73	0.76
Europe			98.5	99.0	99.2	0.99	0.99	1.00
Latin America and Carib.			86.6	89.7	91.0	0.98	0.98	0.99
North America			99.9	99.9	99.8	0.99	0.99	0.99
Other countries/areas			73.4	78.6	80.5	0.77	0.81	0.82
World			76.2	82.1	83.9	0.85	0.89	0.90

Illiteracy is more prevalent among women. Out of 31 countries in the region where data are available for the years between 2005 and 2007, only the Republic of Korea and Singapore had literacy rates favouring women, while 10 other countries were within the range considered to represent gender parity—a GPI between 0.97 and 1.03. The situation was worst in South and South-West Asia, where only Maldives was within the parity range. By far the lowest GPI was in Afghanistan—at 0.29 in 2000. Bhutan, India, Pakistan and Nepal also had low ratios, around 0.6-0.7, but these are slowly rising as overall literacy improves. All North and Central Asian countries have reached gender parity in literacy.

The interpretation of literacy data is not always straightforward. Literacy statistics are usually drawn from censuses or household surveys that rely on self-assessments or third-party reporting, or use educational attainment as a proxy. Generally these overestimate both literacy and functional literacy. A number of literacy assessment surveys have attempted to measure skills and literacy profiles in a more comprehensive manner, such as the Literacy Assessment Surveys in Cambodia (1999) and in the Lao People's Democratic Republic (2001).

Another important measure of educational outcome is the percentage of children enrolled in primary school who reach the final primary grade. Among 29 Asia-Pacific countries for which data are available, 20 countries had 80% of students reaching the last grade. However, in a number of countries the proportion reaching the final grade is lower: in Bangladesh in 2005, 55%; in Cambodia in 2006, 55%; in the Lao People's Democratic Republic in 2006, 62%; in Nepal in 2007, 62%; in India in 2005, 66%; in Pakistan in 2004, 70%; in Myanmar in 2006, 73%; in the Philippines in 2006, 73%; and in Fiji in 2006, 76%. On the other hand, all the countries in North and Central Asia had more than 90% students reaching the last grade of the primary level. Elsewhere other countries had equally high retention numbers: Brunei Darussalam; Hong Kong, China; Indonesia, the Republic of Korea; Samoa; Sri Lanka; Tonga; Turkey; and Viet Nam.

In line with a change in the official MDG indicator, the definition of survival rate at primary level has changed from that used in the previous *Yearbook*. Now we measure survival to the last grade of primary level rather than the survival to grade 5. This definition will thus offer compatible information for all countries, regardless of the official duration of primary school.